Professors'
Guide

to

GETTING
GOOD
GRADES
IN COLLEGE

"I have long believed that the way we do Life is the way we do college. Therefore, this isn't just a book about how to get better grades. It's a book about how to do Life better. Hugely important! Very well done. I recommend it—to everyone."
—Richard N. Bolles, author, *What Color Is Your Parachute?* 9 million copies in print (revised annually)

"This book is not just about getting good grades; it's about getting the most out of your educational opportunities in college. Sure, if you follow the clear and insightful tips that Jacobs and Hyman offer, you will get good grades; but you will also maximize your learning."
—Robert J. Gross, dean, Swarthmore College

"Every student who uses the tips and techniques in this volume is virtually guaranteed a grade increase. Even better, these students will understand their subject matter better, will have better relationships with their professors and fellow students, and will have a set of skills and proficiencies to carry forward to graduate school and professional careers."
—Sharon J. Hamilton, director, Indiana University Faculty Colloquium on Excellence in Teaching

"Jeremy and Lynn are passionate teachers with an uncanny genius for making the most complex ideas clear and assessible to developing minds. This book on surviving college is a gift to college students everywhere."
—Kim Kowsky, former student of Jeremy, freelance writer/editor, and former *Los Angeles Times* staff writer

"Brutally honest, lots of fun, and refreshingly specific, Jacobs and Hyman scatter the clouds of confusion surrounding the pursuit of great grades and chart a practical path to success. Reading *Professors' Guide to Getting Good Grades in College* is the ethical equivalent of using x-ray vision on your teacher's mind."
—Troy Smythe, former student of Lynn, and principal, Smythe Consulting—Adaptive Museum Planning and Research

"*Professors' Guide to Getting Good Grades in College* deserves an A+. Who better for students to receive advice from on how to be successful students than from professors themselves?"
—Eric R. White, executive director, Division of Undergraduate Studies and associate dean for advising, The Pennsylvania State University

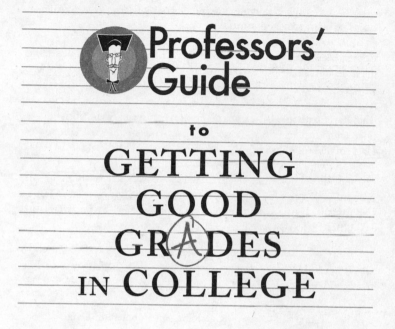

Professors' Guide

to

GETTING GOOD GRADES IN COLLEGE

Lynn F. Jacobs and **Jeremy S. Hyman**

Collins
An Imprint of HarperCollinsPublishers

FIRST EDITION

Book design by Nicola Ferguson with Corrine O'Neill

Library of Congress Cataloging-in-Publication Data

Jacobs, Lynn F.
    Professors' guide to getting good grades in college / Lynn F. Jacobs
and Jeremy S. Hyman.
        p.   cm.
    Includes index.
    ISBN-13: 978-0-06-087908-2
    ISBN-10: 0-06-087908-4
    1. Study skills.   2. Grading and marking (Students).   3. College
student orientation.   I. Hyman, Jeremy S.   II. Title.

LB2395.J29   2006
378.1'703281—dc22   2006041248

---

07 08 09 10 WBC/RRD 10 9 8 7 6

This book is dedicated to you, the student.
May everything you touch turn into an

Ⓐ

Lynn and Jeremy

CONTENTS

PART 3   THE EXAM

PART 4   THE PAPER

PART 5   THE LAST MONTH

INTRODUCTION

This is a book about getting good grades in college. You won't find anything here about getting along with your roommate, making friends and hooking up, balancing study time and party time, or how to do your own laundry. No, it's all about grades. A's—and how to get them. You'd be amazed how many times students come to us at the end of the semester saying, "I'd do anything for an A." By then, of course, it's too late. The course is over. The die has been cast. But for you, A's are available for the asking. Or at least the reading.

This book is written wholly by professors. The folks who give the grades, and who know what it takes to get good grades. Throughout the book we'll be giving you a behind-the-scenes look at all aspects of the grading process. We'll show you how the college grading system is set up, what items count (and don't count) toward your grade. We'll reveal what the professor is thinking as he or she constructs the course, gives the lectures, and assigns and grades your tests and papers. And we'll expose how professors size up students in discussion sections, office hours, and even from e-mail communications. Once you understand how the professor is thinking about all the activities that affect your grade, you'll be able to do them better. This will put you on the fast track for getting excellent grades in your courses.

You might have thought this information was available elsewhere—for example, from the professor or TA teaching your course. Think again. Professors today are much shorter on time than they were even five years ago, as a result of booming enrollments (often without additional instructional staff). At UCLA, for example, the super-obscure Medieval Philosophy course used to have 30 students; now there are 200 (plus a wait-list of 50). Also, professors are often very reluctant to talk about how grades are given and what you can do to get good grades. Some think that it's "unbecoming" for professors to talk about grades, or that students are already too obsessed with grades (so why rev them up more?). Others just find it too unpleasant to talk about grades and want instead to stick to discussion of the Javanese gamelan, Pavlovian conditioning, protease inhibitors, or whatever the course is about.

That's why this book will be an enormous help to you in your quest for that most important of college prizes—the golden A. The *Professors' Guide* is brimming over with high-value, authoritative tips, techniques, strategies, and methods. Instead of just telling you to sit in the front of the lecture hall and pay attention, we'll show you how to use the verbal and behavioral cues of the professor to construct an excellent set of lecture notes. Instead of just telling you to study hard for tests, we'll show you how to anticipate questions, triage your time, and make full use of the course resources. Instead of just telling you to go see the professor (or TA) when you're having trouble, we'll tell you what to say (and what not to say) when you're there, so that you can get all the help you need to reel in that truly excellent grade.

All the tips in this book are practical and easy to use. In each case we'll tell you exactly what to do, and how to do it. And the tips are time-savers, too. Nowhere do we direct you to work harder, or to try harder, or to do all the course activities for their own sake. You might *find* yourself working harder, or trying harder, or even (gasp!) going to all the classes. But it'll be because once you understand how the various activities fit into your getting good grades, you'll be more *motivated* to do them. And you'll enjoy them more, to boot.

Trust us, the tips in the *Professors' Guide* really do work. Between the two of us, we have taught over 10,000 students (that's right, ten thousand) at a total of eight different universities—Lynn at the University of Arkansas, Vanderbilt, Cal State Northridge, University of Redlands, and NYU; Jeremy at Arkansas, MIT, UCLA, and Princeton. We've seen thousands of students move from B's to A's—and hundreds from C's to A's. We know how *they* did it, and we'll show you how *you* can do it, too!

But there's another reason the tips, techniques, strategies, and methods of this book will work. It's because once you understand the *mechanisms* of grading—what counts (and what doesn't) for the grade, how professors make up and grade tests and papers, and how professors are willing to help you get good grades—you'll be able to do the things that *produce* good grades. You'll do the right things, you'll do them right, and you'll get good grades. It's that simple.

*   *   *

This book is arranged around the five "grade-bearing moments" of the academic semester. Much like the load- or weight-bearing walls of a house, these are the times and activities of the semester that bear the full weight of your grade in the course. And which, if done correctly, will result in an A for you at the end of the course.

If you choose to read the book from cover to cover, you'll find the grade-bearing moments arranged in the order of the typical academic semester (or quarter, if your school has those). You'll begin with (1) *The Start* (in which you'll dispel common myths, learn how professors grade, and pick courses with an eye to grades); you'll move on to (2) *The Class* (in which you'll learn how to drop and add courses, how to take excellent lecture notes, and how to structure a typical week); you'll proceed through (3) *The Exam* (in which you'll learn how to best prepare for, take, and go over your test); you'll tackle (4) *The Paper* (in which you'll learn how to understand the assignment, how to think out and write both analytical and research papers, and, most important, how to enlist the help of the professor); and finally you'll arrive at (5) *The Last Month* of the semester (where you'll learn to surmount the hazards of this major grade-bearing moment, and how to prepare for, and take, that all-important final exam).

But maybe you're more pressed for time, or you're picking up this book well into the semester. No problem. Each of the 15 chapters stands on its own and will work without any of the others. So if you find yourself with lecture notes that could fit on a postcard (and still leave room for the address), have a look at our tips in Chapter 5 for taking excellent notes. If you're quaking in your boots about the upcoming Organic Chemistry exam, take a peek at Chapter 8 on how to ace exams by adjusting your attitudes. Never done a research paper and think the Internet is just for "social networking"? Page through Chapter 11 and get fully up-to-date tips about electronic databases and e-journals.

It's the last month of the semester and you're getting a grade that's less than you hoped for? Check out Chapter 14 and take advantage of our techniques for moving from a B to an A (or a C to a B).

You'll be happy to hear that the book is fast-paced and enjoyable throughout. No 600-page textbook here. No professors holed up behind the podium, droning on endlessly about who-knows-what.

Each of our chapters begins with a punchy introduction and ends with a summary *review session*. Some chapters include engaging "top 10" lists or snarky "do's and don't's" tables. But if that weren't enough, we've included a series of "boxes" designed to instruct, motivate, and (yes) entertain. Some of these—*Professors' Perspective, Lynn Remembers When . . ., In Our Humble Opinion,* and *Visiting Professor*—are designed to show how the professor is thinking about some aspect of grading. Other boxes—*Extra Pointer*s, *4-Star Tip,* and *Wann'an A?*—provide additional, high-value tips tailored to specific or unusual circumstances. *You Can Do It!* boxes offer motivational advice, *College Speak* boxes explain special college jargon, and *It Happened Once . . .* boxes relate interesting anecdotes that illuminate one or another dimension of the grading process. (Future editions of *Professors' Guide* will add *Student Input* boxes, featuring the best tips and stories submitted to our **www.professorsguide.com** Web site. Send one in!)

The *Professors' Guide to Getting Good Grades in College* is designed to demystify the process of grading, and help college students get truly excellent grades. *You,* for example. If this book helps you to get *one* A (where otherwise you would have gotten a B or, worse yet, a C), it will have been a success. If you succeed in getting a *couple* of A's, the book will have been a barn-burning success. But if on December 22 or

May 22 (or whenever it is at your school), you open up your e-mail, only to find a note from the Registrar saying you got *all* A's—well, wouldn't that be sweet!

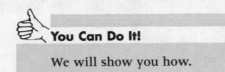

You Can Do It!

We will show you how.

PART 1

THE START

CHAPTER ONE

10 Common Myths about Grades in College

Most students arrive at college with lots of baggage. We're not talking about suitcases full of clothes or boxes stuffed with Urban Outfitters swag. We're talking mental baggage that often includes lots of myths about college grades. As professors, we hear these myths all the time—usually when a student hasn't done as well on a paper or exam as he or she had hoped. These myths are bad news. They're not just false, they can be destructive. They can stand in the way of your getting good grades. Mythbusters to the rescue!

POP QUIZ

Myths and Facts about College Grades

ARRRRGH! Wouldn't you know it? A pop quiz—and on the very first day. You know the drill. Take out your number 2 pencil. Consider the 12 statements below and blacken the boxes to indicate which of these statements are true and which false. Be sure to blacken the boxes completely. Set your timer for 5 minutes and begin your work . . . NOW.

	TRUE	FALSE
1. Most college students don't care about grades.	☐	☑
2. The goal of college is to earn a diploma, not to get good grades.	☑	☐
3. College is usually easier than high school.	☐	☑
4. Effort counts in college grading.	☐	☑
5. Attendance counts in college grading.	☐	☑
6. Being nice to the professor will help your grade.	☐	☑
7. Grades are most often just the professor's opinion.	☑	☐
8. Some students just can't get good grades.	☑	☐
9. Professors don't really care what grades students get.	☑	☐
10. Professors usually explain how to get good grades.	☐	☑
11. The professor often isn't the one who gives the grades.	☑	☐
12. In some courses, practically any number grade can equal any letter grade.	☐	☑

The answer key to this quiz is given later on. And it's printed upside down, just like answers always are. So be on the lookout for it. (Keeps you at the edge of your seat, doesn't it?) In the meantime, here's the scoop on the myths—and why they *are* myths.

MYTH # 1: "It's Bad to Be a Grade-Grubber"

We hear the terms all the time: dweeb, nerd, geek. Or worse: butt-kisser, brown-noser, grade-grubber, teacher's pet. All of these names, of course, are used to put down students who are just trying too hard to get good grades— students who are ill-adjusted and undersocialized, and will go to practically any length to get that A. That's why many college students are ashamed to admit they care about grades at all. And why we see such students coming to office hours with their tails between their legs, sheepish about displaying even the slightest interest in grades.

But let's face it, grades are the *currency* of college. They're the money; they're what counts. No one should feel apologetic about wanting to get good grades. According to the U.S. Census Bureau (and hey, these guys would know), a college diploma can cash in at $1 million in increased lifetime income compared with a high school diploma. A high GPA can get you into top professional schools and higher-paying jobs. Just like a high batting average is a sure ticket to a lucrative contract in the major leagues, or an excellent bottom line is a sure route to a large end-of-year bonus for a CEO. Do you ever see baseball players embarrassed by good stats? Could you imagine a CEO apologizing to stockholders for the company's record earnings? So why should *you* be embarrassed about wanting good grades—the measure of success at *your* job?

Clearly you have a healthy respect for grades (otherwise you wouldn't be reading this book). And there's absolutely nothing wrong with that. Indeed, there's something absolutely right with that. So don't let some little voice in your head (or your friend's head) tell you that only geeks and nerds care about grades. That voice could keep you from getting where you want to go. Turn that voice off and stand up for your desire to get good grades.

MYTH #2: "Why Try to Get Good Grades? All I Need Is That Piece of Paper"

Some students enter college caring only about getting the diploma. They figure that grades don't really matter. All they have to do is pass their courses. For them, college is about killing time till they can dress up in their cap and gown and pick up their ticket to those high-paying jobs. Why worry about grades when you can become president of the USA (or candidate for president) with a gentleman's C?

Here's why. Nowadays there are more people out there with a piece of paper than ever before. Not to mention highly skilled workers in places like India to which jobs can be outsourced. So for plenty of employers, a college diploma isn't enough. They want to know your GPA. They want to know that you have good skills and work habits. They want to hire someone who stands out from the pack.

Just aiming for the diploma will take you out of the running for internships, merit-based scholarships, and honors that could dress up your résumé. It will take you out of the running for many postgraduate educational opportunities. More important, you could end up with a diploma and no real skills or knowledge needed for the next step.

But most important, not aiming for good grades—and the success and achievement they reflect—can suck the life out of your college experience. You slog through the courses, never really trying, never really investing the effort that could actually be the enjoyment of taking all those courses. You never find your passion, you never engage in much of anything—for all you care about is the end, the piece of paper.

MYTH #3: "College? This Is Going to Be a Cakewalk"

Some of the newest arrivals on campus are certain that college is going to be a breeze. They plan to party all night, play lots of poker, go to football games—and cut classes, skip the reading, leave the papers till the last minute. And still get good grades. After all, these recent-comers pulled straight A's in high school—and easily. And they've heard it doesn't take much effort to get a decent grade in this college. "After all," they think, "how hard could it be—they admitted me!"

But consider the math. Most of the students who go on from high school to college are in the top percentages of their high school class. Take UCLA, where Jeremy went to graduate school. The University of California accepts only students in the top 12 percent of their high school classes. Once these students get to the UC, a full 88 percent of them are going to drop below the rank level they had in high school.

Or consider the fact that most of the freshmen at the UC have never gotten a C in high school, but they are going to be taking college classes in which about 30 or 40 percent of the students will get a C. When you consider these numbers, you see that the odds are against students' being able to achieve the same grade level in college as they did in high school. It's just the math. You won't be able to beat these odds if you come in expecting a cakewalk.

MYTH #4: "E Is for Effort"

This is one of the most common of all myths about college grading. That effort counts. It's hard *not* to fall for this one, because in elementary and secondary school, effort is strongly rewarded. High school grades often take into account how hard a student has worked, how much he or she tried, how much time was put into the task. A high school paper often gets a higher grade just for being longer or for having used more sources. And then there's the good old extra credit, the ultimate expression of the high school equation "more effort = better grade."

In college, though, this equation does not hold true. That's because professors are grading how good your essay is, how well you solve the problem, or how well you perform on the test—in short, the end result. They are not grading your effort.

Professors usually don't even have a reliable basis on which to measure effort or to compare your effort against anyone else's. Usually the class is simply too big for the professor to notice how hard you worked; and even in smaller classes or discussion sections, it's hard for the professor or TA to come up with an objective way of measuring how hard you tried. And besides, when you're in college, you're in the big leagues. Should a world-class orchestra include cellists who miss half the notes but try really hard? Would an expert surgeon be judged on how hard he or she tried to complete the operation, regardless of its outcome? Why should professors give good grades to someone who tries really hard but still doesn't get the point of the assignment or the correct answers on the test?

Professors' Perspective

Old myths can be hard to shed. You know you're still thinking that effort counts if you find yourself saying or thinking:

How could I have gotten a C, given that . . .

- I wrote five drafts of this paper?
- I worked on this for 3 (4, 5 . . . 100) days?
- I studied twice as much as my roommate, but he/she got a better grade?
- I bought the lecture notes, went to the review session, and even met with the TA before the test?
- I consulted 10 print sources and a dozen Web sites?
- I put all the information on flash cards and copied over my notes?

These (and dozens of variants) are lines that professors hear all the time. When we hear them, we express sympathy or encouragement, we offer erudite reasons for our grading decision. But what we're really thinking is "That may be, but your paper or test still wasn't all that hot."

MYTH #5: "A Is for Attendance"

As long as there are elementary and high schools that give prizes for perfect attendance, there will be students arriving at college thinking that attendance is the key to getting good grades. These students will show up for every class, rain or shine, even the ones right before Thanksgiving break that everybody else is cutting. These students are never even a minute late. Maybe they even deliberately plunk themselves right in the front row so the professor can see their shining faces the minute he or she starts class. Everything seems to be going just great—that is, until the first test or first paper is returned. That's when professors are likely to find their offices invaded by students outraged by their less than stellar grades. "But, Professor," these students proclaim indignantly, "how could I have gotten such a bad grade? I never missed a *single class*."

Don't get us wrong here. We believe that students should attend all the classes. Later in the book we'll get up on our soapbox and expound on the importance of going to class. But however valuable attendance may be, it's simply not the case that attendance will net you a good grade in a college course. That's because professors do not consider attendance at all when grading an exam or paper. Some professors will allot some small portion of the final grade to attendance or, more likely, class participation (perhaps 10 or 20 percent). But usually the bulk of the grade (well over 70 percent) is based on how well you perform on papers and tests.

That's because college grading focuses on *product*, not process. You are graded on the quality of the work you produce, not on how you've managed to produce it. Attendance can be part of the process, and an important part at that. But no amount of attendance will count much if it doesn't result in a good product—a bang-up paper or exam.

MYTH #6: "If Only I Kiss Up Enough . . ."

And while we're at it, you may be thinking that if effort doesn't count and if attendance is nada also, maybe you can at least charm your way to an A. After all, professors are human, too, and they'd surely welcome a little flattery. "Now if only I can manage to do it subtly—in such a way that the professor barely notices what's going on, and my friends don't notice it enough to brand me a brown-noser—then I'm home free."

We've seen it. We know it in all its varieties. There's the "ask a question after every class" version, the "talk to the professor about all the extra reading you've been doing" version, the "tell the professor how much you love the field and all the other courses you've taken" version, and the painfully obvious "tell the professor how great a class it is and how great a professor he/she is."

Yeah, we know sucking up when we see it. To tell you the truth, we actually kind of like it. But it doesn't make any difference in our grading. Some professors, ourselves included, even grade blind—not that we close our eyes, but rather we grade without knowing whose paper or exam it is (so as not to be influenced by how much we like, or don't like, the student). And, of course, many professors have absolutely no idea who is in their class, anyway (so it makes no difference if they read the name before they grade).

Still, there is a lot to be said for forging good relations with your professors. Some of the most important "4-star" tips in this book will be about ways to do this. We'll show you how to get your professor to spill the beans and give you information that will open the doors to the kingdom of good grades. These secrets—among the best we have to offer in this book—are not to be confused with simply sucking up, however.

MYTH #7: "Grades Are 100 Percent Subjective . . ."

Some students are convinced that grades are completely subjective—that there's simply no fact of the matter about whether a paper is an A or a B (or a C or a D), that grades are just the professor's opinion, and often a bad one at that. Such students get a C on a paper and wonder (sometimes to themselves and sometimes out loud—very loud), Why did I get a C? Did the professor just not like me, or was he or she just too stupid to understand my view? Or was the professor just mad at me because I said something that he or she didn't agree with? Or was it because I missed class a few weeks ago? Or that he or she didn't like the way I looked, or dressed, or seemed? The list of possible explanations can go on and on.

This myth needs to be busted ASAP. This myth is not only false but destructive. You simply can't make any progress toward good grades if you think the grade is all in the mind of the professor. This book is going to show you many steps you can take to get good grades. But the first of these is realizing that grading isn't merely subjective, but is based on the *quality* of your work. That there is a method to this madness of grading—and in this book you will learn what it is.

When professors grade, they aren't assigning grades arbitrarily, but are judging the students' work against certain standards. If you were to read a full set of graded papers in a particular class, you would see a consistency in the professor's grading judgments. When we read a paper that has left out half the answer, that has completely missed the point of what we've been teaching for three weeks, or that says one thing on one page and the opposite on another, we don't for a moment think that the paper might get an A. On the other hand, when we read a paper that rises above the level of quality of most of the pack—because it has a deeper understanding of the materials or more original insights into the issue—we know right away that this is A-level work. In both cases it's because there's an objective standard we are applying—not an exact template that every paper is matched up against, but a basic understanding of what would constitute a good answer to the question asked. And what, of course, would constitute a less good answer.

We'll have lots more juicy information on how professors grade coming up soon, but for now recognize that grading is a *system*, not just an opinion. And once you figure out this system, you'll be able to work it to your advantage.

MYTH #8: "I'll Never Get Good Grades. I'm Just Not a Good Student"

Some students have internalized the belief that they aren't good students. They figure they're just not the *type* of student to ever get really good grades. That somehow it's written in the stars (or in their genes) that they just aren't going to do well. Other students convince themselves that they lack certain skills or are not good at particular subjects. They can't write well, they're bad at languages, math, science— you name it. Maybe they haven't gotten such good grades in the past. Maybe their friends or family have told them they're no good at that sort of thing.

We believe, however, that *every* student is capable of getting A's in college. And we are not alone. *The fact that you have been admitted to college shows that the college also believes that you can do well.* This book is designed to give you the tools to get good grades in college, regardless of how you have done in the past.

We have seen students move from D's to A's and from being on academic probation to being top students. We know a student, now working on her third master's degree, whose family told her she was too stupid to go to college. (Maybe one day she'll even progress to a Ph.D.—or two.)

Of course, setting out to do well in college when you think you aren't a good student—and when you don't have a lot of confidence in your abilities—can be scary business. It might seem easier just to resign yourself to doing adequately. Then at least you won't be disappointed. But why settle for less than the best? Why cheat yourself? Especially since we'll give you the information—and the encouragement and the support—to get past the fear and start getting good grades. Once you discard this self-defeating (and sometimes self-fulfilling) myth, you have already opened the path to getting good grades.

MYTH #9: "The Professor Could Care Less What Grade I Get"

Many students feel a sense of alienation about all things having to do with grades. They sense (they think) that the professor doesn't care at all what grade any student receives—that the professor is a sort of grade-dispensing automaton who spends 5 or 10 minutes skimming each piece of work, slaps an A or B or C on it, and then moves on to the next. All without emotion. All without caring. And then the student thinks, "Why should *I* care at all what grade *I* get in this stupid course? If the professor—whose course it is—isn't at all interested in my grade (and the achievement it might conceivably represent), then why should I—a mere passive observer—have any interest at all in whether I get an A or a B (or a C or a D)?"

Turns out, this picture is false. The fact is that professors feel really good about giving out A's, and really bad about giving out C's and D's (about B's it's hard to feel much, one way or the other). This might amaze you, but lots of professors truly care about their students. Many have forgone higher-paying jobs in order to go into teaching. Most love their fields and are extremely eager for students to learn and love it, too. And sometimes a professor has worked really hard with some particular student and is really eager for him or her to succeed.

But even the professors who don't care that much about their students care about the grades their students get. That's because professors feel a sense of personal failure when assigning a bad grade. A bad grade is a sign that the professor has not succeeded in teaching that student. Even if they try to blame the student—claiming (and believing) that two thirds of the responsibility is the student's—most professors giving out bad grades can't avoid the feeling that somehow they themselves did a bad job.

So, far from not caring at all, most professors care a good deal about their students, or at least about the grades that their students get. They are eager for their students to do well—if not for the sake of the students, then for the sake of their own egos.

MYTH #10: "The Professor Will Tell Me All I Need to Know to Get an A"

You would think that those caring instructors of your courses would give you full information about how to get good grades in their classes. Surely that's the one thing no professor would fail to cover in detail. For if there's one thing the professor knows, it's how he or she assigns the grades. Especially since, as we now know, professors really do want their students to do well.

But life is just not that simple.

There are as many reasons as there are professors why students are never given full information about grades in college courses. One especially common reason is a lack of time. Professors tend to be obsessed about not having enough time to cover the material. Many fall behind on the syllabus and end up guilt-ridden, having left out chunks of valuable content they had planned to teach. Other professors figure, "This is college. The students should already know how to get good grades." And some professors strongly believe that part of the learning experience itself is for students to figure out for themselves how to get good grades.

(These reasons have always struck us as bizarre. Couldn't the professor have planned the lectures efficiently enough to salvage 20 minutes to explain what to do on a paper or exam? How should a college student know how to do a college-level paper or exam if no one has taken the time to explain it carefully and in detail? Finally, what could be more strange than to expect people to learn how to do something well only by letting them figure it all out for themselves?)

Professors'
Perspective

Having given some official reasons why professors don't tell you all you need to know, we can now turn to the real reasons. Professors dread prolonged discussion about grades in the classroom setting. Many would prefer having all their teeth pulled than talk about grades in class. Talking about grading is tense. Telling people publicly how they will be evaluated is an inherently unpleasant activity. And saying too much about grading can come back to bite the professor, when a student thinks he or she has followed the recipe but the paper or test hasn't turned out all that well. Worst of all, talking about grading (especially after the work has been returned) runs the risk of a public outburst from an upset student, or a challenge or confrontation from an angry, hostile, or sobbing student. So is it really so surprising that the last thing some professors want to do is slowly and carefully lay out what you need to do to get an A?

Luckily for you, from the safety of our study we are here to give you the full scoop on how to get good grades in college—the insider information about how grades are really given and what you can do to get the very best grades. Hang on—we are about to go where most professors fear to tread!

Review Session

Students come to college with a wide variety of myths and preconceptions about college grading. Some think that only misfits care about grades, or that grades don't really matter, or that it's very easy to get good grades. Others think that they'll get really good grades if only they expend a lot of effort, or make it to all the classes, or kiss up to the professor enough. Still others have doubts about the grading system itself or their own ability to function in it. Grades are completely subjective or arbitrary, they think. And in any case, I'm just not a good student, so I'm not going to do well. And finally, there are some who transfer responsibility to the professor, figuring he or she doesn't care, so why should I, or he or she will tell me all I need to know about grades, so why should I try to figure it out.

All these myths stand in the way of getting good grades. When you think that grades are unimportant, or that what counts in college is just what counted in high school; when you adopt a defeatist or fatalistic attitude, or think that what's most important is what the professor feels or publicly offers up about grades, you rob yourself of the chance to get good grades. Before you've even taken the first test or written the first paper. But when you realize that college grading is a rational system, and that there is a step-by-step method you can use to master this system, you empower yourself, you take matters into your own hands, you control your own fate. And you get A's—as you will see.

Thought we forgot the pop quiz? Are you kidding? Here are the answers, upside down as promised.

Answer key:

1-10 are false; 11 and 12 are true (we'll see why in the next chapter).

(Unhappy with your grade? Look ahead to the end of Chapter 9 for info on when to dispute a grade.)

CHAPTER TWO

How Do Professors Grade, Anyway?

At the heart of college grading lies one central moment—the time when the professor picks up your piece of work, reads it through, and decides upon the grade. But what actually goes on here? College students never know because this activity is conducted in total secrecy. Always. Professors go to great lengths never to be caught in the act. But now for the first time we lift the curtain on the college grading process. We show you what goes on behind the scenes—in the hidden recesses of the professor's office. Once you understand how college grades are given—and what lurks in the mind of the professor as he or she doles out the grade—you'll know how to prepare your work in such a way as to get the best possible grades. Are you sitting down? Because your knees may buckle as we reveal that . . .

FACT #1: The Grader Can Be Someone You've Never Seen Before

Sure, the professor gave out the assignment. And sure, when the assignment was handed back, the professor talked to the class about how well they had (or hadn't) done on

Jeremy Remembers When . . .

When I was an undergraduate at the University of Michigan, I took five advanced courses with the same philosophy professor. The professor always provided lengthy and detailed, even typed, comments on my written work—often quite positive. He even went so far as to propose topics for my senior honors thesis—advice that went far beyond the call of duty. So before leaving for graduate school, I went to personally thank him for the years of thoughtful comments on my essays. "Don't thank me," he responded, "thank the guy in the brown jacket who sat in back of you in class." "Who was that?" I asked. "A grad student who's been working for some time as my grader—he's the one who wrote all those comments." It turned out that the professor had never once read my papers or exams in any of the five courses. And had never breathed a word about his hidden, dirty secret. I felt ripped off. And I still do.

the test or paper. But that doesn't mean the professor actually graded the assignment. Think about it for a moment. There's no way a professor teaching 100, 200, 300, or more students in one class is going to be able to grade all those papers him or herself. And why should he (or she), when in many universities there's a stable of eager graduate students ready to take on those tall stacks of papers?

The grader of your course could be a grad student whose sole job is to grade the papers and tests, and who is never identified or introduced to the students in the course. Or, in courses that have both lecture and discussion (or section or quiz or lab) meetings, the grader could be the teaching assistant who leads your discussion section. In some courses, different parts of an exam could be parceled out to different TAs to grade; sometimes in a math test, for example, each problem is graded by a different TA.

Later in this book, we'll show you how to determine who the grader is, and we'll explain why it is essential for you to locate (and if possible meet) the one who actually grades your work. For the time being, notice that things aren't always what they seem: the one who lectures isn't always the one who grades.

COLLEGE SPEAK

A TA is a student, usually a graduate student, working as a teaching assistant. Typically he or she holds sections, grades papers, and meets with students to help them with their work. In some schools these are called TFs (teaching fellows), GAs (graduate assistants), section leaders, preceptors, section men (or women), or a variety of other names. But regardless of the name, the result is the same—the professor is, in effect, subcontracting, or outsourcing, the grading to someone else.

If you're lucky enough to attend a small college or a school without a graduate program, your professor might actually deign to grade your work. But even then, you're not likely to be spared this next startling revelation, that . . .

FACT #2: Grading Goes on Only Very Episodically

As you come to class and discussion day after day, week after week, you might think that the professor (or grader) is constantly assessing and evaluating you and all the other students. Noticing who comes to class (and who doesn't), who talks in section (and who doesn't), who comes to office hours (and who doesn't)—all these, you might think, are grist for the grading mill that is constantly running. Five days a week. 15 weeks a semester. 3 semesters a year (if summer school is included).

But the truth is that the professor or grader is grading you only a few times in the semester. That's because the grading in a college course is pretty much clustered into a few times—usually about three or four per semester—when the grader is occupied with your tests or papers. The rest of the time the professor or grader isn't thinking much about the grade at all.

Don't be fooled by some of the busywork that goes by in lots of courses. Some courses have regular homework, but it's not usually collected or graded (or if it is, it's often just for practice). Some classes have quizzes to test if you've done the reading or memorized some facts. Most of the time all the quizzes and homework in a class count only for about as much as 10 minutes of the midterm when it comes to the final grade in the course.

Don't get us wrong. We're not telling you to skip the quizzes or not do the homework. What we are telling you is that the real time to sweat is when you have to deal with the three or four key pieces of graded work. We've seen students tear their hair out over a quiz that counts 5 percent, while blowing off a research paper that counts 30 percent of the grade. How much sense does that make?

★★★★☆ **4-Star Tip**

If you're fresh from high school, don't be fooled by the fact that not much is happening on the grade front for the first four or five weeks of the semester. This doesn't mean that you can just blow off the class until the first graded piece comes around (the single most common—and most costly—mistake new arrivals make). In Part 2, you'll learn all the things you need to do to queue up for excellent grades—things like picking courses right, going to all the classes and taking excellent class notes while you're there, and preparing properly and taking stock week-by-week. For the time being, though, you need only to *plan* to hit the ground running. Something only one in 10 first-year students really do.

If it weren't enough that grading goes on only once in a while, you'll be appalled to know that . . .

FACT #3: The Grading Act Is Often Over Almost Before It Starts

How long do you really think it takes your professor to read and grade your paper or exam? Do you imagine he or she spends hours poring over your search for meaning, savoring all its elegant turns of phrase and brilliant nuances? Or is it more like an hour? Half an hour? Twenty minutes? Actually, the answer is: none of the above.

The truth is, your work is in and out of the grader's hands usually in about 10 minutes, 15 minutes tops. When Jeremy was at the University of California, for example, each TA was told to allot one—and only one—hour of grading per student per course. Since many of the courses had four graded pieces of work, this averaged out to at most 15 minutes per graded item (even less if you count the time spent entering the grades, consulting with the professor, and so on).

You may think that your work deserves more time and attention. Maybe so, but no one, not even a TA or grader, has lots of spare time at a university. And even if he or she did, he wouldn't invest it in grading—in teaching or research perhaps, but in the dastardly act of grading, never. So the wheels of grading have to move quickly. Hey, this is America, the land of productivity and efficiency.

Later we'll see what steps can be taken to capture the imagination of—and unlock the A's from—a grader reading 50 pieces of work for 10 or 15 minutes each. But for now just realize that the grading act, even the best grading act, rarely lasts more than a few minutes.

The worst of it is, though, that even working at breakneck speed, your grader is doing something totally unexpected to your paper or exam. Take a look deeper behind the curtain and you'll see, to your horror, that . . .

FACT # 4: Your Work Is Going to Be Scrutinized under a Microscope

You may not be used to having your work taken all that seriously. Maybe you don't think your teachers notice if you use a term improperly, or if you keep repeating yourself, or if you talk about stuff you don't really understand. Maybe your teachers don't notice if you never really get around to answering the question or if you just dance around the topic. Maybe they're happy if you say anything at all coherent or

simply fill the pages. If this is the sort of thing you're used
to, then wake up. You're in for a surprise when you hit college.

Graders at college, whether full-fledged professors or
grad students, are professionals. Very professional. And very
experienced, too. Often they've graded thousands—nay,
tens of thousands of exams and papers. As a result they pay
very careful attention to the work they're given. They put
your work under the microscope of their intellect and
knowledge. They're primed to find errors, on guard for any
signs that your work isn't up to snuff. You're not going to be
able to slip anything by them.

One thing that sometimes surprises beginning students is
the degree to which graders write detailed comments on
their work. What you handed in as a set of clean white
sheets with elegant black type is handed back bathed in red
(or, if the professor is sensitive, and doesn't want to seem
too critical, green or purple) ink, hardly a square inch of
your paper free of scribbles. These comments aren't just
general points addressed to the class as a whole, but are
custom-tailored to what you yourself have said, argued, or
proved.

As we'll see later in the book, these comments, depressing as they can seem at first glance, are actually a gold mine.
Because your grader is often a big expert (or at least an advanced graduate student) who is taking the time to engage
with your ideas, to seriously consider what you are thinking
about the issue or problem. And, typically, he or she will offer
suggestions about what additional points to consider, about
how you could have done better, and about how to improve
in the future. In short, truly individualized attention—the
single best moment of college.

For the time being, be aware just that your work is going
to be evaluated far more carefully than you might be used

to, and that (at least in the best case) you'll receive far more feedback than you might ever have wanted. Serious business, this grading stuff. But if that weren't enough . . .

FACT # 5: Your Work Is Graded against a Standard

Anyone who has graded for more than a semester knows the great importance of setting up a standard before embarking on reading a set of tests or papers. What is the main point of the assignment, and what components might reasonably be included in the good answer? What methodology, or type of argument or analysis, is appropriate for this piece of work? These—and many similar—questions parade before the mind of any at-all-experienced grader, before he or she has picked up even a single paper from the stack.

To some degree, the features selected in this standard—and the weighting given to each—are up to the grader him or herself. Some graders will place more emphasis on overall coverage of a topic or issue; others prefer specificity and detail. Some graders are particularly concerned with proper use of specific terminology or jargon in the field, while others care only about the clarity of the exposition. Some will favor a more creative and original approach, while others will want the student to stick more closely to the material covered in lecture or in some assigned reading or article.

But most values are going to be shared by virtually every college grader. So, sorry to say, you'll never find a grader who will reward major errors of fact or interpretation, or key omissions, B.S., grandstanding, or unclarity or vagueness at any major point.

Visitors

Visiting Professor

Ed McCann, University of Southern California

A student once asked me before an exam, "Do you want us to give you our own ideas or just regurgitate what you taught us?" I, of course, answered, "You should regurgitate. But keep in mind, I've read a lot of exams and I'm a real connoisseur of regurgitation. I can distinguish the kind that is well processed and fully digested, from the kind that just comes right up."

Still, all this talk of standards and measurements against standards might be leaving you cold. So to give you an idea of what a standard might look like (and how it might be applied), we reveal here for the first time—and in living color—the standard *we ourselves* use when grading papers. Have a look:

The Truth about Grading: How Lynn and Jeremy Grade

Before reading any piece of work we set up 3 "levels"—basic, substandard, and excellent.

Level 1: Basic, meeting expectations (the B grade)

We start by forming certain basic expectations. Nothing fancy. Nothing embellished. Just the basics. For us, an adequate essay needs to (1) do the whole task, that is, answer all the questions asked; (2) display knowledge of the lecture and reading materials relevant to the question; and (3) approach the material with the sort of analysis and methodology used in our fields (and employed in our lectures).

An essay that meets these expectations in a basically good, if "bare-bones," way will get an 80 (or a low B–). (Other, stricter professors might give such an essay a 78 (or C+), while super-nice professors might even give it an 83 (or low B).) As we read through the essay, we simply compare the essay with this bare-bones standard, making constant adjustments in the grade as the essay exceeds or falls short of the model.

Level 2: Substandard, not meeting expectations (the C grade)

The C essay is one that does not meet the bare-bones standard we have set. Usually, there is a clear point (or series of points) at which the essay goes off the cliff, drops out of the B range. We have been known to let out audible gasps at these moments. Often these points are *mistakes*, which, at least for us, are glaring errors that no good treatment of the material could contain. Equally common is the *omission*—simply leaving out some point that we consider essential to the good answer or, worse yet, failing to answer one (or more) of the questions posed. Still another C maker is confusion, vagueness, or irrelevance. A piece of work that started well becomes unclear or hard to understand, or material is offered that seems to have no bearing on the question asked. (Much more on all of these later, in Chapter 13.)

Regardless of the cause, the piece of work has failed to meet the basic expectations—the minimum standard for the good paper or test. And we, reading the paper or test, feel irritated or annoyed or put off (sometimes even a little sad). "How could our beautifully delivered lectures, and our carefully constructed readings, and all the help we've offered up in office hours and review sessions, culminate in such a mediocre piece of work? C it is!—No doubt about it. On to the next one in the stack."

Level 3: Excellent, exceeding expectations (the A grade)

On a happier note, essays that soar significantly above the bare-bones standard are going to get an A. These essays distinguish themselves by going beyond the minimum of what is being asked. They can do so in several ways. Some show a deeper understanding of the material, perhaps revealing a greater awareness of relations between ideas or points being considered in the question. Other A essays are unusually complete: they bring in data that others haven't thought to mention, or they provide fuller examples to support the points made. Still others have more original insights into the material, or explain the material more clearly than is usually done.

In every case we're left with a feeling of elation. "This student did a great job—he or she really understood what we were trying to teach." A true success, for student and professor alike.

By now you should be bowled over by how fair and rational our grading system is. *We* grade against absolute standards (though ones we've formed ourselves). But there are other professors who do the unthinkable. Unlike us . . .

FACT #6: Some Professors Grade by Ranking One Student's Work against Another's

In some courses grades aren't based on an absolute standard but are relative. This is called "grading on a curve." The way this works is that the professor assigns a numbered grade to each student's test or paper, but then assigns a letter grade to each number, based on the distribution of numbers in the class. The top 20 percent of the tests—whatever numbers they have on them—might get A's. The next 30 percent get B's of various stripes. The next 40 percent—those are the C's. And the bottom 10 percent get D's and F's.

Of course, the professor can manipulate this system to generate as many of each grade as he or she wants. But often courses are curved to produce what's called a "bell curve": most of the students will receive grades in the B–/C+ area, with progressively fewer students receiving grades as you move farther (in both directions) away from this point. If you graph the grade distribution, it will start low with the number of A's, then rise to a peak at B–/C+, and then fall back down symmetrically to a low number of D's and F's. There is no one set system of curving that all professors who curve will use, though there may be established practices within particular colleges or departments.

What all this means in practice is that you have no idea of how you did just by seeing your score. You need to know how your score rated in relation to all the other scores. So an 80 on the first exam in Chemistry 101 could be worth an A or a C, depending on how your classmates did on that test.

To confuse matters even more, some professors will not only curve the individual tests and/or papers, but will curve the *cumulative* point total for the course, so that you will not know exactly where you stand in the class until all the point totals are in. This is the worst of all, for here you have no idea how you are doing, and how much time and effort you need to invest in this particular course.

Professors'
Perspective

You might be wondering why on earth any professor would think up a crazy idea like curving the grades. Well, some professors just think it's the best way to achieve fairness. No matter which semester you're taking a given course, no matter which TA you get, the grade distribution is going to be the same. The professor made it that way. Other professors have the more radical idea that grades have no absolute meaning at all. On their view, a B+ is just a report that you have done better than, say, 70 percent of the other students enrolled in that class.

But for some professors there's a really different reason for curving—fear. Theirs. They're afraid that their exam might be too hard and everyone will do badly. Or that their exam might be too easy and everyone will do well. With a curve, they can guarantee that no matter who is in the class, no matter what kind of exam or assignment they give, the final grade roster will be picture-perfect—at least from the professor's point of view.

Some students take great comfort in hearing that a class will be curved. To them curving means that no matter how badly the class does, the grades will be curved up. These students apparently haven't had the fun experience of taking a class where the grades get curved down.

`SOAP` _IN OUR HUMBLE OPINION . . ._

Whatever the professor's motivation might be, we ourselves think curving is a bad idea. For one thing, it pits student against student, fostering a spirit of competition that we think is unhealthy in a university. For another, it sets an absolute limit on how many A's (or B's or C's) are available in a given course. Even if everyone in the class does excellent work, the number of A's given out isn't going to go up. Finally, the curving system usually assumes that there are the same number of excellent students in a given course, semester after semester, year after year. But what happens if the year you take the course, there are, for some reason, better students—some of whom are being unfairly denied the A's that they have, by the quality of their work, earned?

Obviously, we've never been big fans of "curving the grade." But the phenomenon exists. And now you know.

Visiting Professor
Thomas Wieckowski, Arcadia University

A couple of years ago, I was an associate dean of another college. We had a midterm drop rule. That is, a student could drop only up to the end of the sixth week of the term. Five P.M. Friday, no exceptions. On Monday afternoon of the seventh week, my secretary ushered a dozen agitated students into my office. They told me they were all flunking their Statistics class. No surprise there. I told them sternly that they were a day late and couldn't drop now. They replied that they didn't really want to drop if it could be avoided. They explained that the professor believed in grading on a curve and that half the class should flunk. The half that was flunking had dropped, themselves, by the deadline on the previous Friday afternoon. The remaining

students—who had all been passing on Friday—
showed up for class on Monday to find that the professor had recalculated the curve. Now, half of the
members of the reduced class were magically flunking.
These students in front of me were the new flunkees. I
tried to be appropriately noncommittal—academic freedom of the professor and all of that—but the next day, I
took the professor out to lunch at the faculty club and
we had a man-to-man talk about grading and curves.
The chocolate cake for dessert must have done the trick.

There's just one revelation left within the dark recesses
of the grading process. Can you handle just one more? Our
final secret is that . . .

FACT #7: In Some Schools Most of the Grades Are A's and B's (But That's about to Change)

Everyone going to college knows the grading scale. A (or in
some schools, 90–100 percent) is for excellent; B (or 80–89)
is for good; C (or 70–79) is for fair; D (or 60–69) is for poor,
if passing; finally F (or 59 percent or below) is for failing.
But what you might not know is that at some schools the
most common grade is a B; and D's and F's are almost never
given, at least to students who've bothered to hand in all
the work.

At colleges such as Harvard, Columbia, Duke, and
Pomona, roughly half the grades are A's, with the remainder
mostly B's and a few C's. And at the University of Illinois, a
large state university with a more diverse, and less selective,
student body, more than 40 percent of the grades are A's, and

only about 13 percent are C's. (These figures are courtesy of Duke statistician Stuart Rojstaczer, operator of the wonderful Web site *www.GradeInflation.com*.)

You might think that your ship has come in, that if the grades are half A's and half B's—or even a third A's, a third B's, and a third C's—your chances of getting a good grade are pretty good. And that you've squandered your $15.95 investment in buying this book. Well, first, whether 40 percent or 30 percent or 10 percent of the students in your class are getting C's, it's bad news if *you're* one of the ones getting a C. And, of course, just because the school average is 3.09 doesn't mean that in every course the average is going to be better than a B (especially since upper-division courses and courses for majors are going to pull up the school average).

Don't think you're the only one who knows these stats. Businesses and graduate schools—the ones to whom your GPA really matters—are only too aware of which schools have high grades. They understand that when a college has an average GPA of B, a B student is just average; and a C student, well below the norm.

And what happens if you're going to a school where the average grade is 2.86, or 2.8, or 2.74? At these schools many students are going to be getting C's—and what goes on elsewhere is of no consequence to them. (Statistics for many schools are available at *www.GradeInflation.com*. Check your school. And at some schools the average grade is made public. Ask your Registrar.)

But regardless of what college you're attending, the real news is that the gravy train is about to end. In spring of 2005 industry leader Princeton University announced a plan to limit the number of A's assigned in any given course to 35 percent of the total (previously as many as 50 percent of the students had been getting A's). In other words, a net diminution of a third of the A's. So, many of the A's will have to be pushed down to B's, and one can imagine a systematic "deflation" where some of the B's are pushed down to C's. Anecdotal evidence suggests that Harvard and Duke University, as well, are pursuing informal plans to get the grades down and/or to limit the number of students graduating with honors.

It's reasonable to suppose that other universities will follow suit. Universities are like lemmings: they slavishly follow one another, especially when the big lemmings are moving quickly. And it's not hard to see what the fuss is about. The American—indeed the global—workplace is demanding well-trained college students, ones who know how to think and write, analyze and communicate, not ones who got an easy A and were barely in the top half of their (A-laden) class.

The Truth about Grade Inflation

Lately there's been lots of talk in the media about grade inflation. Grades have been going up, it's said, even faster than housing prices on the coasts. What is the truth?

Have a look:

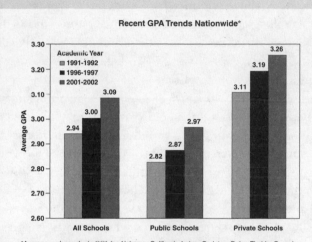

Recent GPA Trends Nationwide*

Academic Year
- 1991-1992
- 1996-1997
- 2001-2002

Y-axis: Average GPA (2.60 to 3.30)

All Schools: 2.94, 3.00, 3.09
Public Schools: 2.82, 2.87, 2.97
Private Schools: 3.11, 3.19, 3.26

*Average undergraduate GPA for Alabama, California-Irvine, Carleton, Duke, Florida, Georgia Tech, Hampden-Sydney, Harvard, Harvey Mudd, Nebraska-Kearney, North Carolina-Chapel Hill, North Carolina-Greensboro, Northern Michigan, Pomona, Princeton, Purdue, Texas, University of Washington, Utah, Wheaton (Illinois), Winthrop, and Wisconsin-La Crosse. Note that inclusion in the average does not imply that an institution has significant inflation. Data on GPAs for each institution can be found at the bottom of this web page. Institutions comprising this average were chosen strictly because they have either published their data or have sent their data to the author on GPA trends over the last 11 years. (www.gradeinflation.com)

Last update Mar. 17, 2003

For these 22 colleges, then, grades are going up at a rate of .15 point a decade. In other words, what was slightly less than a B average grade 10 years ago (2.94) is now slightly more than a B grade (3.09). Some inflation, yes, but not the runaway inflation that's reported in the press. And for you—who are going to be in college for four, five, or six years—it's not likely to make a difference (especially if your school doesn't even have pluses and minuses).

Professors'
Perspective

None of the research we've read indicates the reason(s) for the upward slope in grades over the last few decades. But many times we ourselves have felt the pressure to inflate the grades we give. One source of pressure is the consumer mentality about college and grades. College is unbelievably expensive, and college buyers—students and, more important, parents—feel they should get a return on their investment, i.e., good grades. More personally pressing is our dependence on getting good student evaluations. To our chairs and deans (that is, our bosses), bad numbers often mean bad teaching (that is, bad work performance). And bad teaching often means slower promotion and smaller raises. Finally, the constant grade disputes wear one down after a certain number of years. "How much does it really matter, whether a student gets a B on this paper—or whether you just bump it up to a B+, when asked?"

Of course, we steadfastly resist these pressures—always. But they're there—always.

Review Session

The curtain rises to reveal—a veritable zoo. The zoo that is college grading. Sometimes the grading is done by a total stranger (at least to you)—and, in any case, only once in a while, and at blazing speed. Often the grading is done with great intensity and focus, sometimes against some fixed (though unknown to you) standard; at other times, comparatively—against other students. Sometimes the grade distribution is pretty high—unless at your school it isn't—but that's about to change as there's a contraction in the number of good grades available.

But relax. There's a method to this madness. You can tame the beast. In Chapter 1, you put aside the myths that stand in the way of getting good grades. Now you've learned the who, when, how, and what of the grading process. You're ready to begin your quest for the coveted A in earnest. It begins with the next step—picking courses with an eye to grades.

CHAPTER THREE

FAQs about Picking Courses with an Eye to Grades

Your hunt for the A begins before the semester has even started—with your picking of courses. In front of you lies an online catalog whose size rivals only the results of a Google search (or if your school is still in the Dark Ages, a "course book" or "racing form" or "time schedule" as big as the L.A. phone book). Surely, without too much sweat you'll be able to find four or five selections that interest you, satisfy your requirements, or at least fit your schedule. But wait. What will the effects of your various choices be on what really interests you (if you are minimally rational and still reading this book)—getting good grades? And therein lies the rub. How to know? How to pick? You've got questions. We've got answers.

Q. If my goal is to get A's, wouldn't it be a good strategy to take all easy courses?
A. *No.*

Every college has its share of really easy courses. These are classes where barely any work is required, where you can pretty much sit in the lectures (or not) and be guaranteed an A (surely a B). At some schools they are called "guts" (for reasons that no one really knows); at other colleges, "micks" (short for Mickey Mouse). Info about which courses are easy spreads around campus faster than computer viruses. Don't tell anybody we told you, but at the University of Arkansas "Death and Dying" is a much-sought-after course. We hear that "Human Relationships and Sexuality," at the University of Vermont, is similarly regarded—though with a more appealing subject matter. At many colleges, rumor has it, courses on children's literature (or "kiddy lit") are a prized way to fulfill English requirements without having to read anything above a second-grade level. And then there are "Rocks for Jocks" (Geology 101), "Chemistry for the Consumer in the 21st Century" (Chem 119), and "Mathematics of Powered Flight" (Math 104).

Wouldn't it be a good strategy to make a directed search for such courses, then lard up your schedule with them? Well, the truth is, it would—at least in moderation, and in the short term. We have no problem with you taking an easy course every so often, especially if you're interested in the content. Indeed in some cases this is a positively good idea. But the strategy of selecting all easy courses isn't really going to work as a long-term plan for getting good grades in college.

That's because there isn't an unlimited supply of gut courses. There's just no way to avoid difficult courses altogether. Maybe some of the courses you are *required* to take are designed not to be so easy. Strange but true, the faculty teaching these must-take courses actually think there should be some content built in, and some skills imparted to the students held there against their will. And then there's the *major* or concentration. When you start the real work in your major, you'll always need to take advanced courses. If you haven't built up your skills by that point, you are going to find yourself in an unenviable position.

Choosing a course based on its easy reputation can also come back to bite you. Sometimes you'll be the victim of misinformation. Lynn often encounters students who are shocked and dismayed to find that art history—despite being ranked by David Letterman as easier than lunch—is actually a challenging field. Also, sometimes one person's mick is another person's nightmare. Many students tout speech classes as an easy A, but if you're terrified by public speaking, you might not have such an easy time.

 IT HAPPENED ONCE . . .

> When we were in college we knew a classics major named Tom who thought that "Greek and Roman Sports and Recreation" would be a welcome respite from mind-bending courses in classical languages. So he signed right up for it, and was shocked when he ended up with a C—a much lower grade than he usually got. How could this have happened? Well, it wasn't that "Greek and Roman Sports and Recreation" was harder than reading *The Iliad* in ancient Greek. The problem was that Tom (who tended to nerdiness) had no real *interest* in sports or recreation. He was bored out of his gourd and couldn't muster up enough effort to do well (though of course he was fully capable of getting an A). The moral? If you take a course just because it's easy, and slog through it like a zombie, you might end up with a less than stellar grade. And even if you do manage to get an A, what have you really gotten?

Q. Should I avoid like the plague courses that I know will be hard for me?
A. *It depends.*

You certainly shouldn't take courses in which you are in over your head from the get-go. A physics course that required a good dose of calculus wouldn't be a wise selection for someone with only an 11th-grade math background. A sociology course that presupposed a good command of statistics would be a GPA-buster for someone who isn't really good with numbers. And a French course conducted wholly as a conversation between professor and students wouldn't be a wise choice for someone who'd had four years of high school French but whose class never really included much conversation.

Most colleges try to protect their students from making such foolhardy choices. Often there are three or four levels of skills-based courses offered. In physics, for example, you might find *General Physics, Concepts of Modern Physics, Physics I, Fundamentals of Physics,* and *Why the Sky Is Blue: Aspects of the Physical World* (at least if you go to Cornell University). In spite of the almost perfect synonymy of their names, these courses are conducted on quite different levels of difficulty, they presuppose quite different skills, and they occupy quite different places in an ongoing program in physics. Our advice? Make your selection carefully when confronted with a series of such "tiered" courses. Get the facts. If you're not 100 percent sure which course best fits your needs or background, ask a professor or an adviser.

Another way colleges aim to protect you from yourself is by imposing prerequisites for courses that need significant background. But it's not uncommon to find courses that should have prerequisites, but don't. Maybe the professor didn't think to put them in, maybe when the course listing was approved in 1950 they didn't think of it, or maybe the department just wants to build up big enrollments. And at

COLLEGE SPEAK

A prerequisite (or "prereq") is a course whose completion is required (usually with a passing grade) as a
prior condition for taking some other course. Often an
intro course is a prerequisite for taking an advanced
(or upper-division) course in that field. And sometimes there is a prescribed sequence of courses, for example, World History 101, 102, 201, and 202, in
which the earlier courses in the series have to be
taken before the later ones.

many schools it's pretty easy to talk your way around the
prerequisites by feigning an interest in the subject, by sweet-
talking the professor, or by claiming you've completed
courses equivalent to the prerequisite. Sometimes the computer doesn't even enforce whatever prerequisites there are.

Our recommendation: tread with caution. Don't buffalo
your way into a course for which you lack the prerequisites
unless you are very certain you'll be able to handle the
course. And (as we'll see in the next chapter) if you find yourself in a course that should have a prerequisite, but doesn't,
bail out. The minute you know you're taking on water.

All this notwithstanding, there are going to be courses
that cover material important for your career goals—or for a
possible (or actual) major. You should not shy away from
these courses, even if they might be hard for you.

These courses can be some of the most exhilarating and
motivating experiences of your life. In his freshman year at
Michigan, Jeremy was advised to take an upper-division
course in the history of 17th- and 18th-century philosophy—
a quite challenging course that involved much reading of primary sources, oral presentations in section, and a very

rigorous instructor. This course so inspired Jeremy that many years later he transferred to UCLA graduate school to study this same material (though on a higher level) and with this very same instructor (who hadn't become any less rigorous in the meantime). Lynn, after her sophomore year, was one of three undergraduates who ventured into an advanced art history seminar filled mainly with grad students. All three of the undergraduates had to work their butts off, but they all did well, and all three today are successful art historians. None of them will ever forget the excitement of taking a course that was harder than anything they'd ever taken, but taught them more than they had ever dreamed a course could.

Moral? When contemplating a difficult course, look before you leap. But consider leaping.

Q. Should I consider taking a course with a professor who's reputed to be a very hard grader?
A. *This is a tough one.*

The college grapevine is rife with inaccuracies and errors. Sometimes your source is unreliable—one student who did badly in a course in which everybody else did just fine. Even when you have multiple and confirming sources, you may in fact be dealing with a professor who is rigorous and gives challenging lectures and assignments, but in the end just grades in line with the school average. All bark and no bite (at least not that much bite).

When you come right down to it, it's pretty hard for students to really know which are the professors who actually give out grades that fall significantly below the school's typical grade distribution. One time Jeremy found scrawled in big red letters on the elevator wall "Goddam Hyman C-Giver." Any student seeing that graffito would have inferred that Jeremy was in the habit of giving lots of C's (not to mention being an altogether reprehensible individual).

But in fact, all it meant was that the artist had gotten a C. Nothing less, and nothing more.

Still, there's no question that unreasonably tough graders exist (or at least graders tougher than their compatriots). So if you have good evidence that a certain professor is indeed a hard-ass grader, you need to think carefully before proceeding.

✓⁺ EXTRA POINTERS

There are a variety of ways—some obvious, others less obvious—of finding out which are the super-tough graders. Try some of these:

- Look for a posted grade sheet from the previous semester. Just make sure that it's for the same course—and the same professor.
- Ask another professor in the same department whether you're adequately prepared for—and can "expect to do well" in—the course in question.
- Ask friends you trust—lots of them—what experiences they had with the grading in that course.
- Go to *www.ratemyprofessors.com*. Once you get past the snarky comments, and the disgusting (though not sexist) "chili peppers," you can often find information about grading that is broadly correct.
- Look at a syllabus for the course—online, or posted on a bulletin board, or available from the department office. Sometimes the professor offers clues that the course will be graded really hard—an actual curve, a statement of really tough standards, or a list of requirements undoable by any mere mortal.

So suppose you know that Professor So-and-So is a very tough grader. Should you just pass him or her by? Not necessarily.

If Mr. or Ms. So-and-So is only one of a number of instructors teaching an introductory course, by all means go for the grade. Pick an easier grader. And if the course in question is one you are considering just out of general interest, punt. Leave the heavy lifting to others. (More on this shortly.)

But if you need the course, or if it's very important to your program or your educational goals, don't disqualify the course simply because the professor is a tough grader. A hard teacher—like a good exercise trainer or a rigorous diet coach—can push you to limits you never thought possible. These are moments when you truly feel good about yourself, and which shouldn't be cast aside simply because of grade concerns.

One more thing. We believe that the tips and techniques we will teach you later in this book (especially those that explain how to deal face-to-face with the professor in the office hour—see especially Chapter 12) will work particularly well with the hard-grading professor. The tough grader is often also the tough thinker—the one who displays intellectual rigor in his or her own work, and who respects similar tough thinking in his or her student. That tough professor is likely to be especially impressed with the student who takes the time to think hard—and work with the professor in a joint effort. So you, the reader of this book, have an especially good chance to stand out and excel with the professor who usually doesn't give such good grades. Do not fear the hard-grading professor.

Q. **Is it ever a good idea to take a course just because I'm interested in the subject? Isn't getting good grades job 1?**
A. *It's* always *a good idea to take courses out of interest. And if done right, it not only won't hurt your grades, it'll help. Here's how.*

Before lunging, be sure that, whatever you take, you're not in over your head (you've heard that one before). Don't take a course for which you lack the requisite background, or that is really designed for specialists—when you're not one. It might sound fun to take a painting course—after all, you can throw paint on canvas as well as Jackson Pollock, so why not? But in Lynn's department, for example, Painting 1 presupposes previous painting and drawing, and virtually all the students have completed many hours of art classes. A better idea would be the introduction to art studio, where painting (and other media) are taught from the ground up. Here you can have your cake and eat it, too: the fun of painting (and even some training) *plus* the good chance of an A (after all, you've always been good at painting).

So check it out, and if there aren't any obvious counterindications, trust your intuitions and go for it. College is the time to seize the moment. You have more choices and options than you've ever had, and you're not likely to get them again any time soon. So why not sign up for those courses that really interest you? You might really enjoy them.

Enjoyment is good in and of itself, but it also can motivate you to engage with the material and get good grades. You might find that the course you took just out of interest leads to a major in something you love. And you might even have one of those "aha" moments of self-discovery. That course in evolution, cultural anthropology, music history, or cellular biology might lead to a lifelong interest and career! Not to mention the good grades you will get when studying something you actually have a passion for.

And don't worry if you have interests in areas that don't seem at all related. Sometimes the connections become apparent only later. Lynn once had a student who thought his studies in dentistry and ceramics were completely separate— until the day he made casts of teeth in one class and casts of a ceramic sculpture in the other.

Visitors

Visiting Professor

Thomas Lennon, University of Western Ontario

My first couple of years at college (I went to Manhattanville College), I was kind of a yahoo, frat boy type. My main interests were basketball and hanging around bars. I was doing okay at college, but nothing had really clicked. I didn't have a major. I didn't really have any interests in anything that intellectual.

One day I happened to see a poster advertising the Institute of European Studies, which sponsored junior year abroad in several cities, including Paris. And it looked good to me. I was still living at home at the time and this sounded pretty exotic. So I borrowed from the government, saved some money, and put together the $2,500 needed for the program (this was in 1962). It was a total gamble, academically and financially.

From the moment I set foot in the port of La Havre, I felt totally at home—as if I belonged in France. Even though I had only studied one year of high school French. And I felt even more at home at the Sorbonne. Suddenly I became fascinated by the intellectual scene I found in Paris. I started to read French poetry. And, above all, I started to take courses in Philosophy— Descartes, Pascal, and Kant. And that's how I came to the career I have now as a professor of philosophy. I'd been floundering around before, but my career trajectory became a completely straight line from the minute I got off the boat. Plus, I was able to come back fluent in French and gave a speech in French to the Foreign Language Department that knocked the socks off them.

Q. Do I dare take courses with lots of papers, or should I stick to courses that have only tests and quizzes?
A. *Dare away.*

We know that certain students deliberately avoid taking classes that require papers because they think they can't write, or because they don't like to write, or both. Often this isn't such a good strategy—in either the short or the long run. For one thing, consider the pragmatics. If you sign up for courses that have only tests, it's a sure bet that you will find yourself with several tests on one day—not just once, but two or three times in a single semester. Very bad news. Taking courses that include papers will allow you the freedom to stagger your workload and avoid test overload. And you'll be less likely to get bored if you have a variety of activities to work on.

Besides, whether you like it or not, it's really important to graduate college knowing how to write. Good writing requires both clear thinking and the ability to communicate your thoughts intelligibly to others. These skills have tremendous value. Just pick up the *Wall Street Journal* any day of the week, and you're likely to find some big executive (though maybe not The Donald) griping about the difficulty of finding employees who can write and think well.

There's no need to be blown away by courses that require writing. Most professors with a functioning cell in their brains (that is, in our view, most professors) are acutely aware of the need most students have for training in writing skills. That's why professors usually start the course with easier writing assignments, then build up to longer and more difficult papers. Virtually all professors (or their TAs or graders) are willing, indeed happy, to assist students in need of help. Some professors will read drafts and others allow rewrites (sometimes without risk of getting a lower grade— when was the last time *that* happened with a test?).

And if you take advantage of the opportunities to work with your professor in preparing your paper (much more about this later, in Chapter 12), writing papers can actually be a whole lot less tense than taking tests. Instead of sitting in a room with 275 other students, all sweating like pigs, you can write the paper in the comfort of your own room. At your leisure.

Q. Is it a good idea to "max out" my credit hours?
A. *No.*

In the hectic world of the early 21st century, many students go to college in a rush. They want to get out fast. So they supersize their program and take as many courses as they're allowed, or even get permission for an overload: 12 hours. 15 hours. 18 hours. The more, the better.

This isn't a good strategy if you're interested in getting good grades. Doing this will run you ragged. And no matter what courses you take, you'll find yourself overwhelmed with five assignments or tests, all falling due the same week.

And don't be fooled by the workload during the first month of the semester. You'll be doing everything fine for the first two or three weeks, but then, boom!, you get clobbered by the sudden onslaught of papers and tests.

There are no bonus points given for the number of courses you take, and no special points for getting out of college at Mach 3 speed. Remember, product is what counts, not effort (see Chapter 1, Myth #4). That's why we believe it's much better to pick a limited number of courses—the minimum required, which often is four for a full-time student—and concentrate on *them*. This will allow you ample time to study for all of your tests and write all your papers without having to imbibe gallons of Red Bull or take dozens of Vivarins (or worse).

Keep in mind that your grade doesn't depend on how many courses you take each semester, but how well you do in each one. It's well worth spending a little more time in college and coming out with a high GPA.

✓⁺ EXTRA POINTER

"Easy for you to say, Lynn and Jeremy. But I'm short of money. I can't afford the leisurely track you guys suggest. I need to cram those courses in—regardless of what grades I get."

This is a tough one. We admit it. But before running yourself into the ground and guaranteeing poor grades with a ridiculous overload, we encourage you to think creatively. Could you graduate in four (or five) years by taking some introductory, or required, or distribution courses over the summer? If you're paying by the course, is there some nearby community or junior college at which you can take some of the basics at a lower cost? Have you looked into course credit for any AP work you did in high school (thereby reducing your course load)? Does your college offer any "life experience" credits for work you might have done before coming to college? And, if you're really stuck, could you balance your program with more easy courses, at least while you're financially strapped, in the hopes of salvaging a reasonable GPA?

Q. I'm required to take a foreign language. When should I start, and what language would be my best bet for getting good grades?
A. Une très bonne question. *Start ASAP, and take a language you can do.*

The foreign language requirement is the Fear Factor of the college scene. It usually involves many semesters of course work—often four courses taken over a period of two years. And it can be a real GPA-wrecker. Once your grades in a language course start to slide, it's hard to stop the downward spiral. That's because language courses are cumulative, not just within each course but among the whole lot of them. Didn't learn the irregular conjugation of the verb "to have" in course 1? Guess what? People are still going to be "having" things and doing a whole lot more—in courses 2, 3, and 4.

Be sure to get off to a good start in your language courses. One way is to start on the requirement right away. It's going to take a while to complete, since you have to take each course in order and you can't take two at the same time. Rome wasn't built in a day, and language requirements can't be completed in a single semester—not that we haven't seen some students giving this a valiant, but futile, try. Delaying is particularly ill-advised if you are continuing with a language you have studied in high school. To delay is to forget.

Pick your language carefully. If you liked the one you studied in high school and did well in that, stick with it. You already have a leg up and should be able to place out of some of the required courses. (There's a rule of thumb, though not a particularly accurate one, that one year of high school language equals one *semester* of college.) But if you didn't like your high school language and/or screwed up in it, pick *another* language and get a fresh start.

Think twice before embarking on one of the harder languages, such as Russian, Arabic, Chinese, Japanese, or Greek. Don't subject yourself to learning a new alphabet along with new and complex grammatical structures unless you have a really good reason. Such reasons might include: you want to be a Middle Eastern diplomat (good job prospects there); you

plan to go into business in Asia (contact Wal-Mart or Procter &
Gamble for more details); or your mail-order bride or groom
just arrived from Estonia and can't speak a word of English
(you're on your own here). Of course, some majors might require these languages, so that would be another good reason
for taking them.

And once you start your language requirement, by all
means keep going. Starting and stopping, learning and for-
getting, dropping out and repeating courses, are all just going to make this requirement even more of a headache than
it already is. *Languagitus interruptus*—always a very bad idea.

Q. Will I get better grades if I concentrate on
polishing off my distribution requirements, or if
I start scoping out a major?
A. *This is the most frequently asked FAQ. And our
answer is: yes—and yes. And at the same time.*

Huh?

Many students come to college thinking that the first
thing they will do is to get their requirements out of the
way. Get the drudge work done, they think. There'll be
plenty of time for fun later. And sometimes we even hear
parents chiming in, "Do what you're supposed to. Take your
medicine."

There's nothing wrong with polishing off some requirements. After all, you'll have to complete them sooner or
later. But you should never have a semester filled entirely
with courses that you view as chores to be ticked off a list.
Doing that sucks the life out of your college experience.
Once you've had all energy sapped out of you, it's a lot
harder to muster the wherewithal to go to class, to study,
and to do all those things you need to do to suceed in your
classes. Avoid this half-dead syndrome by following this rule:

★★★★☆ **4-Star Tip**

Always include some courses every semester that you really look forward to taking.

A few exciting courses can make some of the duller ones easier to swallow.

The exciting courses you select could well be ones that are part of a chosen major or a potential major. But don't feel like you have to select a major the minute you hit college. Many students arrive at college completely undecided about their major (and stay that way for a number of years)—which in many cases is a good thing. If you're one of these students, relax. There's plenty of time to decide about your major. For now, go ahead and explore some areas of current interest. These are bound to help you zero in on a major, either because they stimulate your interest further or because they totally extinguish it.

Q. What about that "freshman experience" or "getting used to college" course that they're suggesting I take. Will this help my grades?
A. *Yes. Just not directly.*

Recently many schools have instituted *freshman experience,* or *getting used to college,* or *first-year experience* courses. The purpose of these courses (not to be confused with *freshman seminars,* in which you actually study some material) is to help ease the transition from high school to college, and in this way reduce the dreaded and relatively high dropout rate of freshmen at many colleges (sometimes as high as 30 percent). These courses can be optional or mandatory, non-credit-bearing or offered for one, two, or three credits.

Most of these courses are designed to give you practical information and teach various skills (such as study and library skills) that the school thinks you will need to succeed in college. But even more important, these courses aim to provide small group support and mentoring, often by role models you can actually look up to. In fact, it's not that different in concept from what you'd find at a Weight Watchers meeting. (Unfortunately you shouldn't expect to be losing any weight in your first-year experience course.)

These courses are still pretty new additions to the college scene. So far they seem to have a pretty good track record in helping students both stay in and do well in college. So we think that, for most students, it's well worth it to make the freshman experience course part of your freshman program. Even if it doesn't directly help your grade much (how much can a one-credit A count in a sea of other grades anyway?).

One day these courses will use *Professors' Guide* as their textbook. Won't that be great!

What to Do If You're Closed Out of a Course

It often happens, especially at oversubscribed state universities and community colleges, that in spite of your best intentions you're "closed out" of a course. "No more spaces," the computer tells you. "Bye, bye."

There are two strategies of attack. First, and simplest, is to try "off-peak" hours. Like the airlines, college courses fill up in proportion to the desirability of the time. So if you're willing to take a course on a Friday afternoon or a Monday morning—or in some schools virtually any day after 2 P.M.—you're almost guaranteed a place. Same class, different time. And, who knows, you might enjoy the smaller enrollments, or a class after your siesta.

The other possibility is to try to sweet-talk your way into the (only apparently) closed course. When a professor lets extra people into the class, he or she is taking on extra work. What would induce anyone to do *that*? The answer is that many professors are idealists. They want you to learn. They want to teach students who share their enthusiasm for their subject. Many professors have great memories of their own college years. So if the professor sees you as being a lot like he (or she) was, back then, he (or she) will naturally like you, and be much more inclined to give you a break. With this in mind, here are five key things to say, and five things not to say, to increase your chances of getting into that closed course:

THINGS TO SAY	THINGS *NOT* TO SAY
1. "I'm a senior majoring in X, and this is the only chance I have to take this course."	1. "The course I *really* wanted was Y, but there was no room in that one."
2. "I've always wanted to study X, and this course sounds really exciting."	2. "I'm enrolled in the other section of X, but it's at, like, 8:00 A.M."
3. "When I took Intro to X, I did well and really enjoyed it!"	3. "I need something on Tuesdays and Thursdays, and 2:00 P.M. would be really sweet."
4. "I've heard really great things about this course!"	4. "I gotta take a Humanities course, and all the other ones seem, I dunno, kind of less interesting. . . ."

THINGS TO SAY	THINGS *NOT* TO SAY
5. "I've heard really great things about *you*!"	**5.** "I've heard that this course isn't too bad, you know, difficulty-wise."
(*Note:* All of these strategies are best pursued face-to-face, as early during course registration as possible, and with the perkiest attitude you can manage convincingly.)	(*Note:* If you don't see why these are bad things to say, we'd suggest you consider enrolling in Psych 101.)

Review Session

Depending on what courses you pick at this first of the grade-bearing moments, you can be well on your way to getting good grades in college, or you can be erecting impediments (without even knowing that you are) to getting those A's. Like a high-wire tightrope artist (or at least so it sometimes seems), you have to choose between easier and more challenging courses, between the professor who grades easy (if you could even figure out who he or she is) and the killer-grader, between courses that truly interest you and those that only fulfill some or other requirement. All the while making sure the courses you want are really "open" and can be shoe-horned into your schedule.

We counsel balance. A difficult course, or a course that excites a real passion in you, or a course with lots of papers, is often a very fine choice. But don't go nuts—make sure you look before you leap, and never jump into an ocean when you can barely swim. Taking risks is one thing—and often a good thing, especially for the long run—but betting the house is another.

But more than that, we recommend taking responsibility and control. It's your program—*you* have to actually take the courses and fight for the grades in them. Don't let some adviser—who might come equipped with some "standard freshman program"—talk you into taking some courses that don't interest you, that you feel you are not prepared for and cannot do well in, or that simply are not what you came to college to learn. You (or your parents) are paying lots of dough for this program. Select, don't settle!

PART 2

THE CLASS

CHAPTER FOUR

Your Action Plan for the First Week of Classes

The starting gun goes off. Classes start. And you're full of inertia. The inertia of rest. You're still in summer (or winter) vacation mode—sporting a nice tan, wearing cool shades, all mellowed out. Trouble is, what's needed now is the inertia of motion. Because there's real action right from the start—especially for the grade-hungry. Sure, on the surface it seems that not much is going on. What happens in the first class, anyway? The professor comes in, introduces him or herself, hands out the course syllabus, makes a few off-the-cuff remarks, and lets the class go home early. If you skip out, or show up and zone out, what's the big deal? The class hasn't even started. Or has it . . .

Most students arrive at the first class with a rush of emotions. What's this class going to be like? Is the professor going to be interesting? How hard are the tests and papers going to be? Why did I take this course anyway? Who is that dude or babe sitting next to me? Why is it so crowded? And hot? Questions blend into questions. And more questions. And then the professor walks in. The class falls silent—after all, it's the first day.

The professor starts talking. Gives his or her name, rank, and serial number. Says a few words about the subject matter of the course. And what his or her approach is going to be. Gives you a two-page document, and offers a few words by way of explanation. But all that's for later, you think, so you fold up the handout and put it in your notebook. The professor asks if there are any questions. Someone mumbles something about the wait-list. And then you're dismissed. Not so bad, you think. It's taken only 20 minutes. And, thank God, nothing has gone on.

But believe it or not, in those brief moments of the first class amazingly much *has* gone on. For in the professor's introductory remarks, in his or her brief musings on the syllabus, and in your general "impression" of the professor are contained enormously important clues about what the course is going to be like—and whether it's a good idea for you to be taking it. Some of which bear directly on the grade, and most of which will never come your way again.

Most students won't even realize the gold there is to be mined. They are just happy to get out of there and move on to their next mind-numbing first class. But for you, we offer more. Once you understand what's going through the professor's mind in that first class—how he or she thinks about and constructs that first lecture—you'll be able to tease out those hints that will enable you to get an excellent grade. Not to mention actually enjoying the beginning of your study of the new discipline. Seeing how the professor thinks about the subject, seeing how the professor organizes the material in his or her own mind, seeing how the professor views the methodology and tools of his or her discipline—all of these can be objects of genuine pleasure for the inquiring mind. Yours, we mean.

The first week of classes is no time to be in laid-back, vacation mode. It's time to take action. The right action. For best results, follow our simple, three-step plan.

STEP # 1: Be There—and Be On

First things first. In order for any of this to work, you've got to *be* at the first meeting of all your classes. Make it your business to attend, plain and simple. Sure, it would be nice to extend your summer or winter vacation by a day or two. Sure, it would be nice to surf Facebook or MySpace for the third (or eighth or 108th) time that day, searching for your old high school buddy, or better, someone to hook up with. Sure, it would be nice to load up on T-shirts and caps (you've got to look good, after all), or try out that 10 A.M. pizza at the Union (everybody's got to eat). But it's more important that you hit that first class.

Getting to the first class can be harder than it seems. A pretty high percentage of students—even ones with good intentions—don't manage to pull it off. Every semester, without fail, we encounter well-meaning students who were no-shows at our first class. Some got confused about their schedules, others went to the wrong room, and still others spent the hour running around the campus like chickens without heads.

But it's not enough to safely make it to the right place at the right time. You have to be focused and ready to work when you get there. You have to be *on*. This is harder than it might seem. It's really easy to be distracted and disoriented, as well as nervous and scared, on the first day of college. We can relate, because we ourselves get geared up and stressed out at the start of every semester. It's a normal response, at least for us Type A personalities.

Common First-Day Hazards
(and How to Avoid Them)

- **Not knowing where the building is.** (*Remedy:* Download a campus map from the college Web site, and try out the route before the first day of classes. A little extra walking is good for you, anyway.)
- **Not being able to find your classroom in a huge building.** (*Remedy:* Go into any official-looking room in the building—not including the restrooms—and ask. Alternatively, stop and ask anyone in the building who doesn't look like a chicken without a head.)
- **Getting to the classroom listed on your schedule and finding no class, or the wrong class, there.** (*Remedy:* Before heading off to class, check your printed schedule against the most current online schedule for any last-minute changes. These things change faster than airline schedules.)
- **Oversleeping your 8 (or 11 or 12) o'clock class.** (*Remedy:* Catch some ZZZZs the night before. Save the "all-nighters" for later in the semester, when they're really needed.)
- **Being unable to find a parking space.** (*Remedy:* Check out the parking situation before the first day. Why delay finding out that there are not nearly enough spaces, and that you'll be cruising into college in that big blue bus, not your hot black SUV?)

And even a mellow, Type B student may find it difficult to be focused in the alienating environment of some of the larger college courses. You might not be used to classrooms of 400–500 people where the professor is about a football field's length away. In this environment, nothing is easier than to become very passive and just sit back and observe the passing show, without firing up any of your brain cells. Not to mention the fact that we're all trained to be spectators when sitting in large, anonymous auditoriums. In the movie theater, in the concert hall, at the basketball game—sit back, relax, and enjoy the spectacle. They're the performers; you're the *audience*.

You need to fight off this temptation. You need to be proactive. You need to be the barracuda, not the sponge. Because today is the day you're going to get your first really crucial information about grades, from the ones who are responsible for the grades—the professors. Which brings us to step 2 of our first-week action plan. . . .

STEP #2: Decode the Syllabus

In most classes the main activity of the first day is the presentation and discussion of the syllabus. Most universities require faculty to provide a syllabus for their courses, and some schools prescribe in detail what needs to be included on the syllabus (at the University of Arkansas, for example, we need to include a "snow policy"—in spite of the fact that it snows at most two days a year). And while there might be faculty who just hand out the syllabus and say no more, many faculty feel the need to "go over" the syllabus with the students. Just reading through the syllabus is pretty boring, perhaps not for the student but for the instructor (who might well have taught this course a dozen times before). So going over the syllabus quickly morphs into elaboration and explanation of the syllabus—better known as dropping major-league hints about how to do well in the course.

COLLEGE SPEAK

The syllabus is an outline or brief statement of the main points of a course of study. Like the program of a concert, it tells you what the content, structure, and approach of the course are going to be. It is also a sort of contract, often including information about the course requirements and policies.

But aside from sheer boredom, why do professors drop hints? Well, as they read through the syllabus that they haven't necessarily bothered to update, professors think of all sorts of problems that have come up in the past. And which they want to head off at the pass. It's the beginning of the semester, and professors (like any other normal human beings) want the job to go well—which, for them, is for *the class* to go well. Besides, on the first day of the semester they haven't gotten their guard up yet. They haven't fallen back into their "professorial" lecturing mode. They're just talking. So they're willing to dish up much more than they would at any other time of the semester.

The unbelievable irony of the situation is that virtually none of the students even realize what's going on. So instead of writing down the very timely hints the professor is offering about how to ace the course, the students are just sitting back, blithely watching the passing show. If they only knew . . .

★★★★☆ **4-Star Tip**

Take *very* elaborate notes on anything the professor says about the syllabus that isn't already written in it. Write directly on the syllabus (if there's space) or in your notes using the same headings that are on the syllabus. Write down the professor's exact words if possible.

Still, the syllabus and even the professor's "explanation" of the syllabus are often communicated in code—a sort of shared language between professors and those students who've been around the block a few times. That's why even if you read the syllabus very carefully, and listen very closely to the professor's explanation (taking careful notes, of course, since you wouldn't want to neglect a 4-Star Tip), you could still miss some of its true meaning.

So here's a lifeline. To help you see how much grade-bearing information can be contained in the first class meeting, we present an actual syllabus from one of Lynn's courses—complete with a moment-by-moment account of what's going on in Lynn's brain as she talks through the syllabus. Once you see what *this* professor really means—and how many clues about getting a good grade are contained in the very first lecture—you might be able to decipher what the most important professor really means—*your* professor. And you'll never go back to your old way of listening to the first lecture.

In the pages that follow, match the bracketed footnote numbers ([1], [2] . . .) on the left-hand pages with the numbered explanatory comments on the right-hand pages, to see what Lynn is really thinking.

Lynn's Syllabus: What Lynn *Says*

Art History (ARHS) 2923:
ART HISTORY SURVEY II[1]
Dr. Lynn F. Jacobs
575–2000

Office Hours: Tues., Thur. 12:30–1:30 and by appt. [2]

Office: Fine Arts 40

E-mail: ljacobs@uark.edu
Course Web Page: www.ahsurv2.uark.edu

Teaching Assistants: Maria Cho, William Flugel, Bob Gibbs, Henri Maas, and Jill St. Clair.

Course Description and Goals:
This course is the second half of a two semester survey of the history of Western art. The course covers the art of the Renaissance, Baroque, 18th and 19th centuries, and Modern periods. [3] We will focus our attention on many of the best-known artists [4], including Michelangelo, Leonardo da Vinci, Bosch, Rembrandt, Caravaggio, Monet, van Gogh, Cézanne, and Picasso. Some of the issues to be considered include [5]: the development and significance of style, differing subject matters and their meaning in the culture, the role of art in society, and the techniques and materials used in different periods. The course has no prerequisites, but is normally taken in sequence after ARHS 2913.

LYNN'S MOMENT-BY-MOMENT THOUGHTS—WHAT LYNN REALLY MEANS

[1] You're coming in for the second act. Lots of students will have taken the first half of the survey. They'll know the methods and goals of art history. And the kinds of questions that'll be on the tests and papers. We'll move pretty fast—it's a survey. Lots of things for students to keep straight. On the other hand, it's an intro, so you'll only need to know the most central works and the most basic point about each.

[2] I really like students coming to my office hours. When no one comes I'm stuck surfing the Web: who can do their own research when an inquisitive student can pop in at any moment? Appointments?   if I have to. But three-quarters of the time students never show for the time they set. Still, though it's rare, the very best students will come to talk with me before the papers. Not to mention seeing the TAs, who, after all, are doing the grading.

[3] The course is organized into four periods. That's why there are four graded pieces of work. One per period. Capisch?

[4] I'm going to lecture only about the "all-star" artists. While the text talks about lots of minor guys, we're going to skip them in lecture. If they were smart, students would spend less time on the "minor leaguers" when preparing, and more time on the guys who count.

[5] Here's where I tell the students the four things we'll study in each artist. And here's where I telegraph what sorts of issues will be asked about on the tests. Too bad no one's bothering to write any of this down.

What Lynn *Says* (continued)

Each lecture will analyze 4 to 6 works [6]. The emphasis will be on attaining a deep understanding of a small, but representative selection of works, rather than a superficial acquaintance with a larger number. Throughout the course, we will be using the method of *comparative analysis:* we will compare pairs of works of different styles, or different subject matters, and probe how, in each case, the historical context accounts for the artist's stylistic and iconographic choices.

In addition to the lectures, there is a required weekly discussion section. [7] While some attention in the section might be devoted to answering any questions you have about the lecture, the primary activity each week will be the discussion of a pair of articles from the recent journal literature that present opposing interpretations of some work of art. Two students will lead the discussion each week, but all students will be expected to discuss.

Books for Purchase:

Required: M. Stokstad, *Art History,* vol. 2, ISBN: 0130918504. [8]

Recommended: L. Adams, *The Methodologies of Art: An Introduction*, ISBN: 0064302318. Both are available at U Store and, much cheaper, at various online stores.

Preparation and Attendance: *Before* each lecture you should (1) read the assigned pages in Stokstad [9] and (2) study the assigned artworks on the course Web page *www.ahsurv2.uark.edu* memorizing the artist, title, century, period, and country of each. The assignments, along with a brief outline for each lecture, can be found on the "Schedule of Classes and Readings."

What Lynn Really Means (continued)

[6] A good student's lecture notes will be subdivided into four to six headings, one for each work. Comparisons of pairs of works will be clearly marked, with the main point of difference highlighted. The exam questions will be mostly comparisons—one stylistic, the other icono-graphic. It all seems so simple. At least to me.

[7] One-quarter of the class time is in the discussion sections, so 25 percent of the midterm and the final should be about the articles discussed there. Only a fool would miss the sections. Of course, I haven't exactly *said* that, but I did say section was required. So I'm covered.

[8] The required book is as thick as the phone book—but I've set it up so that students will have to read only 18 pages a time. That shouldn't kill them. Recommended reading? That won't fit into the course much, though it's good for students who have too much time on their hands, or—gasp!—students who'd like to learn more about the field.

[9] I'm not starting each lecture from scratch. Too boring. I'm going to assume what I know to be true of only 40 percent of the students—that they've read the text and studied the Web site images before class. And that they know how to spell the name of the artists and their works. Of course, it'd behoove the students to have a look at the "Schedule of Classes" before coming to class—especially the outlines I've been so nice to provide. *That* would surely help them arrange their class notes. But really, it's up to them. I only make suggestions.

What Lynn *Says* (continued)

Regular class attendance is very important since lectures will go beyond the textbook, at times covering works not discussed in the text. **[10]** In addition, the classes will teach the skills of art historical analysis needed for the papers and exams.

Course Requirements: There are 4 graded pieces of work: two short (3–5 pages double-spaced) papers, a midterm, and a final. **[11]** Topics and instructions will be handed out before each of the papers, and there will be a study guide for the midterm and the final. In addition, there will be a review session before the final.

Grading: Each of the short papers will count 15% of the total grade. The midterm will count 25%. There will be a comprehensive final **[12]**, which will count 30%. In addition, 15% of the grade will be devoted to your section presentation, and your participation in discussion of others' presentations **[13]**. Some consideration will be given to sustained improvement. Students may rewrite <u>one</u> of the short papers for grade improvement. **[14]** Late papers are subject to ⅓ grade penalty for each day they are late (except in case of documented serious illness or extreme emergency).

Plagiarism is considered a very serious offense and will be prosecuted in accordance with the University disciplinary code.

What Lynn **Really** *Means* (continued)

[10] Gee, here's a hint if ever I've seen one. Lectures will have new material, not available elsewhere. And I'll teach the methodology of the field. The one they'll need on the papers and tests. And the one not really taught in the text, either.

[11] A good balance—about half papers and half tests. Papers will be comparisons—just like I'm doing each time in lecture. There'll be information before each assignment about how to do that assignment. Instructions, study guides, review sessions. Still, you'd be amazed how many students never make use of the beautiful resources I've prepared for them.

[12] *Comprehensive* final? Now there's a 4-star tip: there'll be a question taking in all four periods. Too bad I forgot to mention this when reading over the syllabus. Oh well, they'll figure it out. They're *students*.

[13] No need to go over this now. Flugel and friends will explain all this at the first section meeting. Flugel? Boy, do I love that name. Flugel, Flugel, Flugel!

[14] Wow, a do-over. And without penalty. You'd think everyone would rewrite. But usually it's only three or four students. Too bad there's no book that shows students how to get good grades in college.

So there you have it. The good, the bad, and the ugly. But it's not enough that you gawk at what goes on in Lynn's mind as she reads through her syllabus, it's important that you guess what's going on in *your* professor's mind as he or she "goes over" the syllabus in *your* course.

You Can Do It!

Decipher the codes of each of *your* syllabuses by carefully writing down—and thinking about—what the professor says as he or she reads through the syllabus. Stay awake. Stay alert. You can do it! Here are the *top 10* things to be on the lookout for.

1. Any information about the structure and organization of the course.
2. Any explanation of prerequisites for the course, or background that would be helpful, if not required.
3. Any tips about what to do before (and during, and after) each lecture.
4. Any hints about what will count, or what will be asked about, on the papers or exams.
5. Any suggestions about extra credit, rewrites, or other ways of improving your grade.
6. Any statement of what will—and will not—be done in section or discussion meetings, or in science or language labs (if any).
7. Any clues about when—and whether—the professor (or TAs) are willing to meet with students to help them do better.
8. Any hidden course rules or policies, and cases in which exceptions can be made.
9. Any discussion of the relative importance of optional course activities—such as extra readings, additional class meetings, or departmental colloquia.
   And finally (and most important):
10. Any hints, no matter how veiled, about just what you can do to get a good grade.

Now that we've completed our decoding of the syllabus—noticing every dropped hint and every turn of phrase—we come to the third—and most immediately grade-related—step in your action plan for the first week of classes. . . .

STEP #3: Drop—and Then Add

This might seem like the hardest step of all. Who wants to change around the schedule that they worked so hard to get all pretty? Many students simply won't consider this. But this is the step that can have the biggest payoff. The decision to get out, and quickly, of a course that for whatever reason is ill selected is one of the best strategies there is to protect your GPA. Who'd have thunk it—an exit strategy from grade-busting courses? Sans penalty.

And it's not hard to do, once you get your mind around it.

★★★★☆ **4-Star Tip**

Immediately drop courses that are not good. Immediately replace them with courses that *are* good.

Most schools have a very established—and very generous—drop/add period. For adds, it might last anywhere from 1 to 3 weeks. For drops, it's usually much longer, maybe even up to 12 weeks (most schools like to give you a really good chance to bail out of courses that aren't working out for you). But we've got a special deal just for you: a *4-day* drop period and a *5-day* add period. Here's what we mean. Decide whether to stay in or drop a class after attending one, or at most two classes. And add any substitute courses (for courses you drop) by the end of the first week of classes.

The reason for this custom schedule is that—in case you hadn't noticed—we really want you to attend the first meeting of all your classes. So, if you drop and immediately replace a course, you'll have a good shot at getting to the first, or at least the second, meeting of the new class.

But how do you know whether to drop a particular course? Well it's an absolute no-brainer that you should drop if you find a course totally incomprehensible. The feeling of being completely lost is a sure sign you need to drop. The quicker, the better. And don't kid yourself that things will improve. It's only going to get worse, because the beginning of every course is as easy as it's going to get. Find something else to take.

Another slam dunk is making a switch when you like the course but not the professor—and there are other instructors offering that same course. Often, especially in large intro courses, there are many professors or TAs teaching different sections of the same course (sometimes listed in the catalog under slightly different numbers). In such a case, anyone with an IQ greater than 2 will take the trouble to drop the less than stellar professor in favor of the new—and possibly much improved—instructor. You'll like the class more, learn better, and get a better grade.

But what about other courses in which it's a closer call whether to stick it out, or cut and run? While it's hard to give a hard-and-fast set of rules for when to drop a course, we think that most students know a bad course when they see it. If only they'd listen to their gut instinct. And once you've seriously opened up for yourself the possibility of dropping—once you see the first week as a trial period with a three-day return guarantee—you'll be more on the lookout for signs of a bad course. And do something about it.

But for those who need instinct-improvement, here are our . . .

6 Best and Worst Reasons for Dropping a Course

BEST REASONS (AND WHY THEY'RE GOOD)

1. I can't understand a word the professor is saying. (The prof won't get clearer and you won't get smarter.)
2. I'm bored to tears. (Imagine what the 20th lecture will be like.)
3. The professor assumes I know lots of stuff I don't actually know. (What the professor assumes is what he or she gets. But not from you.)
4. The course is about something, I guess, just not what I thought it was going to be about. (Foiled expectations, foiled grade.)
5. The course looks like a total waste of time; only a pinhead could stomach this. (Let the other pinheads figure it out.)
6. The course wrecks my program: it was supposed to be my easy course and now it's turning out to be the hardest. (You don't need another A-wrecking course.)

WORST REASONS (AND WHY THEY'RE BAD)

1. I can't hear a word the professor is saying. (Move forward. Plenty of seats in the front.)
2. The subject is new to me. (That's why you're taking the course, remember?)
3. The professor makes us read journal articles, not a textbook. (Wow, an up-to-date course. And at your school!)
4. My boyfriend (or girlfriend) is dropping this course. (You're too codependent. Get over it.)
5. The classroom is too hot. (Don't worry; it'll cool down after other folks drop.)
6. The professor dresses badly. (You think *you* dress like Pierce Brosnan or Nicole Kidman?)

Once you start talking about dropping a course, many voices will try to discourage you. "Don't do it," they'll say; "drop/add will cost you." —But what's $10 when you're paying $10,000? "You'll have to wait in line forever or spend hours getting onto the computer registration system." —But you'll waste even more time studying in a class you can't possibly do well in. "Why change? You have a fine schedule. Why fix it, if it ain't broke?" —It's broken, don't kid yourself. "It's really hard to find replacement courses and get your schedule to work." —Surely there are still some courses available.

So don't listen to these voices. Turn the noise down and do what you need to do. Trust your instincts and go for the drop/add procedure. You'll have gotten out of an insufferable course (instant relief!) and you'll end up with better grades to boot (the long-term payoff!).

What to Do When You're Late to the Party

Smart student. You've actually followed our plan and gotten out of that lousy course. But now you've missed that all-important first lecture of that new class. What to do? Jump off a bridge. (Just kidding.) Actually, you need to:

- **Introduce yourself to the professor after class.** Smile, tell him or her you've just added the class. (It's always good to get to know your professor and look like you're a diligent student.)
- **Request a syllabus** (politely). Ask if there are any tips about the course not included on the syllabus that would be helpful for you to know (then listen and note it all down—you know the drill).

- Never say anything like "I missed the first class—did you do anything?" (This is the royal route to turning off the professor, since it suggests that he or she just wastes the first meeting—which is usually not true anyway.)
- **Borrow the notes** from a student who seems studious—paying rapturous attention and writing nonstop—or who looks like he/she would be a good study partner. Invite him or her to *explain* the notes if things seem to be going well.
- **Relax.** It's only one meeting. There'll be 41 other chances to kill the course.

Review Session

Far from being a chance to extend the summer (or winter) vacation into the academic semester, the first week of classes is the time to get your course schedule in the best possible shape—shape for getting A's, we mean. Think of it as a three-step procedure.

1. **Get to the first meeting of each of your classes—** and in a focused, ready-to-work state. The first lecture usually isn't just a "get-acquainted" session—it's the presentation of the professor's prospectus for the course.

2. **Decode the syllabus, and size up the professor.** Look for any clues about what will count toward the grade in the course; what you're expected to do for the lecture (and discussion section and lab, if any); what the course goals are, and how the course is constructed or organized; and what the course rules are, and whether they are enforced. Then try to assess the professor. Ask yourself: Can I follow what the professor is saying, and is the material presented in a compelling, or at least intelligible, way? Are there prerequisites for the course (either explicit or implicit) and am I on top of them? Can I expect to be able to do the papers and tests, and is the course on the level I expected? Most important, does the course material seem interesting, and is this the kind of professor I can actually imagine learning from?

3. **Decide whether to continue in the course,** or find another. This is the most important step of all. Silence the naysayers, be resolute and forceful, cut and run if the course isn't right for you. Not one in 10 students will do this, but those who do will have taken an important step toward getting good grades that semester. Perhaps you.

CHAPTER FIVE

Top 10 Tips for Taking Excellent Lecture Notes

About 14 minutes into the lecture, it dawns on you. You probably ought to be taking notes. Maybe writing down a few words or phrases from time to time. To relieve the boredom if nothing else. And so that you'll have little reminders when the time comes to study for the test. Surely you should be writing down at least the stuff the professor puts on the board and maybe a couple of words on the handout, too. Most of the time, though, you can sit back and watch the passing show like the rest of your classmates. They're not writing, either, so you're safe.

Far be it from us to disturb this rare moment of tranquility in the hectic world of the 21st century. But if you want to have excellent notes—notes that will actually give you something to study from (not to mention something to keep you awake for the painful 50 minutes of the lecture)—you're going to have to sit up and get busy. Busy writing, that is. If you want to do really well in your courses, you absolutely, positively, without a doubt need to have good notes. Our top 10 tips show you how. . . .

TIP # 1: Create a Document You Can Use Later

In many college courses, especially intro courses, the lecture content is a major subject of study. The professor distills the content into 40 or so well-crafted lectures, paying special attention to selecting those (and only those) points that are most important for you to know. Readings? Mere background. Discussion (or quiz or drill) sections? Most just go over the lecture or, if they have additional content, it too is distilled, this time by the TA. These professors and TAs are great! They do all the processing for you.

But then the lectures take on a life of their own. The lectures become *Lectures*. Unlike your high school teacher, who was content simply to sum up the reading and ask you a couple of questions to make sure you had done it, your college professor thinks he or she knows more than the guy who wrote the textbook. And that part of his or her job is to present his or her own take on the material only touched on in the textbook. Unless of course the professor has written the textbook, in which case he or she takes special pains to have the lectures *go beyond* the textbook (because why would you bother coming to the lecture at all if you get the same information at home?).

And if that weren't enough, once the professor has focused the learning experience on the Lectures, he or she feels free to take exam questions or paper topics straight out of these much-prized Lectures. Imagine that. The professor actually wants a return on his or her investment in these precious Lectures. So how are you going to study all the stuff your professor has been spouting out hour after hour? Need we answer?

You might think there's no need to take notes, because come exam time, you'll remember everything the professor said. Trust us, you won't. We've taught some courses more than 10 times and we still need *our* notes to remember what we're supposed to say. You, hearing this content for the first time, are likely to have forgotten most of it by the day after class, maybe even by lunchtime, and certainly by midterms. So you need to have your own A-1 set of notes that you can study from later in the semester. Not to mention the fact that once in a while a course has "open-book"/"open-note" exams. Here you've really hit the jackpot. You can actually refer back—that is, copy out of—your notes as you take the test. That is, if you have notes worth copying out of.

Your notes are the "textbook" for the lecture content of a course. It's the one textbook you get to write yourself.

The single most important thing about your lecture notes is that they be legible. Clear. Nice and big. You can't study what you can't read. Plus your notes need to be fully comprehensible on their own—a self-contained document that can be understood on its face, without your having to remember what you've long ago forgotten. Our suggestion? Don't just write down short phrases or code words that will mean absolutely nothing when you read them three or four weeks later. Write in almost full sentences or at least in phrases that make sense on their own.

Your notes also need to be complete. This means that they should cover the entire lecture from start to finish. After all, no one tells the professor to refrain from saying something important at the beginning and the end of the lecture (more on these crucial set-up and wind-down times shortly). And your notes need to cover all the lectures given. Studying from an incomplete set of notes is like playing solitaire with a 50-card deck (or smaller, depending on how incomplete your notes really are). You could luck out and get a pretty decent score if you aren't missing a key card. But what if you're missing an ace or a deuce? What happens if the one midterm question is wholly on the lecture that took place before the break (this professor being a somewhat recriminative fellow)? Or on the lecture given the Monday you came in 20 minutes late (you just have bad luck)?

Every day of every semester in every college course a professor is constructing the main questions on a test or paper from a class that some students didn't take notes on. Sometimes hundreds of students. These students have essentially flushed their chances for an A down the toilet.

TIP #2: Plan to Take Notes the Entire Time

Excellent note taking is like swimming. You should come up for air only between strokes, and then only for brief moments. When you're in class you should aim to spend all your efforts on note taking—not checking your e-mail, instant messaging, planning your evening activities, reading the school newspaper, or, worst of all, just sitting back and vegging out. The main activity should be note taking—thinking and writing, thinking and writing, and thinking and writing some more. A sort of aerobics for your hand—and mind.

Yes, we know that not all students are doing this (and that at some schools, in some courses, *very few* students are doing this). But don't be fooled. We're steering you right and they're wrong. Guaranteed. One hundred percent.

In college courses, professors have a mere 40 or so hours in which to cover the entire course content. We, and many of our professor friends, spend lots of time agonizing about how to fit everything in. With the result that our lectures are very condensed and packed with content. Why do you think our lectures start on the stroke of the hour, and go till

Professors' Perspective

Countless times we've seen students coming into an of-fice hour to talk about a not-so-good grade, and when we asked to take a peek at their notes, what we found was a moonscape. A few craters where the student had gotten down what we said—and often quite well, too. But surrounding that, mostly arid, uninhabited waste-land. Places where, for whatever reason, the relevant content simply didn't appear. Why subject yourself to such possible embarrassment? And why queue up for a less than excellent grade? Make sure your lunar rover is roving all the time.

exactly fifty minutes past? And why do you think that in some classes there's barely ever space for a student ques-tion? This jam-packed content needs to make its way into your notes—because, of course, like all professors we firmly (if quixotically) believe that, if we said it in class, you should know it. And the only way all this stuff is going to make it into your head is if it makes its way into your notes. And you know the only way it can get there.

This does not mean that you're supposed to mechani-cally and unthinkingly write down every word or noise that comes out of your professor's mouth. Burps, coughs, or worse—you can skip those. You're not supposed to be pro-ducing a transcript, or sentence-by-sentence rendition of the hour-long lecture. Rather, you're looking to create a reasonably complete record of all the main points the pro-fessor is making—and a pretty good report of how the pro-fessor is making those points.

In the good case, there'll be lots of active thinking going on in your mind as you're constructing the notes. A *dynamic, real-time* process of understanding and recording will be going on. You'll of course be listening to what the professor is saying. But at the same time you'll be using the various bits of information you have about the content (from previous lectures, from previous points in the present lecture, from your reading, from your general knowledge) to place what the professor is saying into some sort of meaningful whole. *No longer are you a mere spectator; you're now a participant.* And you'll be doing all this "on the fly": the professor speaks, you understand, you write it down in your notes. For the whole hour you're actively engaging with and (dare we say) enjoying the lecture. Or, at least, more likely to be enjoying the lecture than you would otherwise have been (sorry, we can't promise you bliss in every class).

Students sometimes ask us how to know what to write into their notes and what to leave out. We think if there's a question, it's best to write *more* rather than less. You can always erase, or just read through points that you later judge secondary. It's much harder to remember and add new content to your notes.

★★★★☆ **4-Star Tip**

Of course each class is different, but we think a good rule of thumb is that you should aim to produce about *a page* of notes for every *15 minutes* of lecture. At least if the professor doesn't ramble too much, or go off on tangents too often. This amounts to about three or four pages for a M/W/F lecture and five or six pages for a T/Th class. (Take corrective action if your notes are coming out significantly shorter.)

What to Do When You Can't
Take Notes Fast Enough

Some students struggle to write fast enough to keep up with the pace of the lecture. Probably the most common question raised in a lecture is "Can you repeat what you just said?" or "Could you repeat reason 3 of the four reasons you've just given?" Some ways to avoid having to ask the sorts of questions that will test your professor's frustration tolerance are:

- Write faster (be the Lance Armstrong of writing).
- Write script, not print (you're trying to get it down, not look pretty, and script was invented just because it's faster).
- Use abbreviations for common words and for names of people, places, and events (just make sure you'll be able later to understand them yourself).
- Familiarize yourself with such names *before* the lecture (often they appear in the reading, on the syllabus, or in a lecture outline).
- Try to use material from the readings, from previous lectures, and from previous points in the present lecture to anticipate what the professor will say next (such "preinformation" is often helpful in understanding faster, hence writing faster).
- Take notes on a tablet PC or laptop, or use a note-taking program such as OneNote, to speed things along (be sure to try before you buy, since students either love or hate electronic note taking).
- Make an appointment early to see your TA or professor and talk honestly about the trouble you're having (they've seen it hundreds of times before and, like your doctor, they'll know what to do).

TIP #3: Capture the Overall Structure of the Lecture

After completeness, the single most important feature of good notes is that they capture the structure of the lecture. It may surprise you, but professors generally have an order and an organization for the lecture. And they follow that structure pretty consistently throughout all their lectures. Once you understand and can recognize a lecture's structure, the lecture will be easier to follow, easier to understand, and, best of all, easier to take notes on.

Have you ever listened to a pop song while tracking verse, chorus, verse, chorus, *bridge*, chorus—the way the songwriters construct these songs? It sounds different once you've understood the structure. And were you aware that even a seemingly free-flowing program like *The Tonight Show, Evening News,* or *The Price Is Right* is really built up from a series of separate segments—all timed in exactly the same way, and standing in the same relation, time after time, without fail? So too, an academic lecture is typically built up from a series of parts. Organized in the same way, and with the same timing—class after class, week after week. When you detect and then follow the *structure* of a lecture, the lecture changes from an aimless drone to an ordered series of points that can be more easily grasped by your mind and recorded in your notes.

We don't mean to suggest that every professor structures the lecture in quite the same way. Professors who tend to be more anal—Lynn, for example—divide the class into neat segments: title, introduction, body of the lecture (itself divided into three subpoints), and conclusion. Jeremy is content to have two-part lectures—the first part always seems to end at 32 past the hour, the second ending at the 50-minute mark (it always takes longer to get out the first major point than the second). Some professors always seem to get through 80 percent of their content in the first 45 minutes, then cram the remaining 20 percent into the last five minutes. And still others always carry over the last few points into the next lecture.

★★★★★ **5-Star Tip**

Figure out—in each of your courses and for each of your professors—how he or she is structuring the lectures. And attend to this structure in all of the lectures, and, most important, use it to generate a really organized set of notes.

But it's not enough to just get down the order of the parts; it's important to capture the structure *within* the parts. The subparts, too, can stand in some relation to each other (and to the whole). Was something used as an *example* of one of the points? Your notes need to indicate what the example was, and what it was supposed to show. Did the professor give a series of reasons that, taken together, *account for* some particular phenomenon? Your notes need not only to enumerate and separate the reasons, but also to capture what the reasons were meant to explain (and how). Did the professor make a claim, present an *objection*, then go on to refute that objection? Your notes need not only to catalog the three different stages, but to reflect what the professor's overall line of argument was. If this substructure doesn't make its way into your notes, you could end up thinking that the professor was asserting something he or she was actually rejecting. And when studying your notes for that paper or exam, you'll see only a series of disconnected—or misconnected—points, not a sustained and directed argument on behalf of some conclusion.

TIP #4: Copy and Make Full Use of Outlines (Wherever They Appear)

Professors often give students outlines of their lectures. Sometimes these outlines are written on the board before class. Sometimes they've been given out on the syllabus or accompanying Schedule of Lectures. Sometimes they're handed out right before class. And sometimes they're unfurled, line by line, as part of a PowerPoint presentation. But wherever these outlines make their appearance, they are a real gift—among the most precious note-taking aids. That's because they lay out the structure of the class, thereby sparing you the mental effort of having to figure it out. Free, no charge. Once you have an outline, it will be a breeze getting your notes to follow the structure of the lecture. The work has been done for you—by the one who set it all up.

Copy all outlines into your notes. If the outlines are available before class, or in those few minutes before class starts, copy them then, so as not to waste class (that is, note-taking) time. Indeed—for true geeks—copy the outlines into your notes before even going to class (if they are available, say, on the syllabus). And once you've copied the outline, use it. Use it to track the progress of the lecture and to shape the structure of your notes. Check off the points as the lecturer goes through them—at least mentally, if not with a pencil.

TIP #5: Make Full Use of the Lecturer's Verbal and Behavioral Cues

Any spoken human communication incorporates verbal cues to enhance meaning. Changes of tone, alterations of speed, use of certain logical connector words, simple pauses and changes in facial expressions—all of these help a speaker make clear to a listener what he or she means. That's why listening to an oral presentation always has meaning that no transcript ever fully captures. "Pay attention to the demeanor of the witness," they tell you when you're on jury duty. "Look for verbal and behavioral cues to determine if the witness is telling the truth."

So too with college lectures. Most professors give many verbal cues to indicate where they are in the structure of the lecture, and how important a particular point is to that structure. You can tell the professor is just getting started when he or she makes grand and artificial-sounding pronouncements, such as: "In our last class we saw . . ."; "In today's lecture we are going to consider . . ."; and "The next major topic for us to look at is . . ." Face it, no one talks that way outside of the college lecture. Which is why the professor is using such marked, and stilted, language. Other professors (those who are very formal, or very worried they are going to forget how the new topic goes) retreat behind the lectern to check their notes whenever they are about to begin a new topic. *Use such verbal and behavioral cues to locate (and record in your notes) the beginnings of major sections of the lecture.*

Once the professor has gotten into the body of the lecture, the most important cues are breaks (again both verbal and behavioral) that slice the lecture into parts. Sometimes you'll hear the professor making some sort of self-conscious statement to mark a transition: "We now turn to the question of . . ."; "Let us now consider . . ."; or "Let's now move on to. . . ." (Funny how often the word "now" appears at crucial turning points, and how the professor always turns the lecture by inviting you to consider something.) Also be on the lookout for phrases that explicitly state the *function* of (and *reason for introducing*) some (sub-) point in the lecture.

Favorites include: "Let's consider an *example* of . . ."; "Now let's look at the *reasons* for . . ."; or "What can *explain* this phenomenon?" And finally, some professors (either consciously or subconsciously) pause for questions from students only at key junctures in the lecture. (These questions function in much the same way as commercials in a television program, giving the professor and students a much-needed break—a chance to catch their breath before moving on to the next segment.)

The conclusion to a lecture is the one part that nearly every student can automatically recognize. Professors are amazed by how alert students are to cues that a lecture is about to end. One has only to say, "To sum up . . ." or "So today we have seen that . . . ," and students suddenly light up. The trouble is, in their haste to get ready to leave, those students often miss those three or four summary sentences that give the piece of the puzzle that was missing—the overall point, or overall structure of the still seemingly disconnected lecture. "Didn't lay it out so clearly in the beginning," the professor thinks. "Well, no damage done, at least I'm getting around to it at the end."

Be alert for the various verbal and behavioral cues that any lecturer uses to get across his or her meaning to the assembled multitude. And make sure that the structure indicated by those cues makes its way into your notes.

TIP #6: Things Written on the Board or on PowerPoint Slides Are Always Abbreviations, and Need Explanation in Your Notes

Professors have one surefire method to get students to take notes. If a professor writes something on the board or puts it on a PowerPoint slide, the students are sure to copy it into their notes. Could there be anything wrong with this picture?

Well, the problem is that all the things the professor writes on the board or on the slide are just abbreviations: code words, code phrases, and even code sentences. It's far too time-consuming for a professor to write more than a few words on a blackboard. And no one wants to put up PowerPoint slides with scores of words. It doesn't look good and, more important, professors don't want to sit around while their students copy 800-word essays. *Whatever professors put on the board or screen is a kind of shorthand that gets fleshed out and explained aloud by the professor.* Always. Every time.

If you just copy what's on the board or on the screen, your notes aren't going to have enough information for you to understand the point the professor was trying to make. You also have to take notes on the professor's verbal explanation. Those words on the screen are very deceptive. They seem so informative, but without the additional explanations they're going to be Greek to you, when it comes time to actually understand them, say in studying for the test.

TIP #7: Be on the Lookout for—and Record the Definitions of—All Technical Terms

Most professors are very concerned that you learn the vocabulary of the field that they are teaching. That's why you will often encounter them using technical terms. In Jeremy's philosophy classes every other word seems to be either "metaphysics" or "epistemology." Lynn can barely get out a sentence that doesn't contain either "style" or "iconography."

COLLEGE SPEAK

A technical term is a word, term, or phrase peculiar to a particular art, science, or occupation. Technical terms constitute a specialized vocabulary for a particular field, so usually only people in that field will use that term in that sense. Even the dictionary doesn't always provide a definition of the term sufficient for someone studying in the field.

When professors use a technical term for the first or second time, they will usually write it on the board and define it. Many students will proceed to write down the word, but not the definition. The result is that they will be able to consult their notes and see the term—perfectly spelled, underlined, and in all capital letters—but without any indication of what that term could possibly mean.

Mind you, the professor is cheerfully thinking, "Now I've explained this term and put it on the board (once or even twice), so now I can go ahead and use it over and over, day after day, without defining it, and the students will surely understand what I'm talking about." Professors have seen, heard, and used these terms for years, so these technical terms are a basic part of the way they think about, and talk about, their field. If you don't get the definition into your notes, though, you could get stuck hearing these words over and over, and just scratching your head. This could go on for an entire semester.

TIP #8: Set Off Any Examples (or Other Illustrative Material) in Your Notes

It's also a good idea to develop (and consistently use) some system for setting off examples—or problems, diagrams, tables, graphs, or test cases—in your notes. Since examples, tables, and the like are not main points, but rather subpoints used to support or illustrate a main point in the lecture, your notes need to *subordinate* the examples to the main point. Lynn likes to set off examples by indenting. Others might use labels or boxes as ways of setting these off. And the truly great note takers can even give headings to these "setoffs" as they're writing them down. But whatever system you use, you should make it so that with just a single glance at your notes you can identify the examples, see their point, and understand their relation to the primary point. Surely your professor included that example, table, problem, or graph with some thought in mind. A good set of notes will make that reason clear.

TIP #9: Do Not Worry about Canonical Note-Taking Formats

Did anyone ever teach you some special format for taking notes? You know, the one where the main headings were Roman numerals, the subheadings capital letters, and the sub-subheadings Arabic numerals? Or were the subheadings small letters? Or were the letters for sub-subsections? And that (i)—is that a letter or a number, anyway?

Don't bother trying to remember these systems, or obsess about using them if you actually happen to remember them. Because these kinds of artificial systems serve virtually no purpose. Who cares how the headings are numbered?

Such systems are also a giant waste of time. You don't want to be trying to figure out if you're at point I.a.2. or III.5.b. while your professor is giving out serious information—which you never wrote down because you were so confused about your numbering system. Usually, a simple 1-2-3 numbering system, with some ways of indicating examples or reasons used to support particular points, is more than sufficient to capture the structure of a lecture. At most, a few numbers with some a's, b's, and c's underneath.

The canonical note-taking system can overly complicate, and distort, what a normal lecture is really like. We've seen students stretching out—indeed, sometimes even adding in—points in an attempt to "fill out" what they take to be the needed structure. On the other hand, canonical note-taking systems can overly simplify, also distorting, the points the professor made. Sometimes students, slavishly addicted to the note-taking systems they learned in 10th grade, carefully enumerate the points of some lecture, but leave out all the connections of thought *between* the different points.

Our conclusion? Blow off those formal outlining systems that force you to shoehorn the lecture content into pre-established boxes (or lines). Think of your professor as the Greyhound bus driver—"Leave the driving to us." Of course, you'll still need to "let your fingers do the walking." But you already knew *that* from tip #2.

TIP #10: Take Notes for Yourself—and for Keeps

Notes are more than just a study aid for tests and papers. They're more than just something you need to use *later on* in the course. Note taking is something that affects your learning in real time. The very act of taking notes is a learning experience in and of itself. When you take notes, your mind has to be active. It has to process the materials the professor is feeding you. It has to discern the structure of the lecture, understand how the examples work, grasp the points being made. All this stuff to do makes the neurons in the brain start firing, and once this happens, whether you know it or not, you are actually learning. The more neurons that fire, the more you learn. The real beauty of note taking is that it's one of the few things in life that gives you both instant and delayed gratification.

So why waste time nodding off, or drifting off in class when you could be learning right then and there?

And if that weren't enough, taking notes starts off the process of internalizing the content of the course. So when it's time to review your notes, you're ready to reinforce what you've already begun learning, not learning from scratch.

The notes you take in class should be notes that you're taking for keeps. Once and for all. No do-overs. No copying over your notes, or typing them up later. And please, no taping the class so that later you can take a second set of notes. All these activities are monumental wastes of time. How could it be better the second time? If you're ever in a good position to generate an excellent set of notes, it's surely when you're maximally engaged, when all the cues are there, and when you're getting the information right from the horse's mouth—that is, in the lecture. Do it right the first time.

One last thing. Lecture isn't the only time you should be taking notes. Really valuable info can be coming your way during more informal class activities, like section meetings, for example. Be sure that content makes its way into your notes, as well.

SOAP **IN OUR HUMBLE OPINION . . .**

At some colleges you can purchase "lecture notes" from the university bookstore. These are usually taken by some "official note taker"—perhaps a graduate student or advanced undergraduate—and are sometimes even "reviewed" by the professor him or herself. Sure it seems it would be good to have a set of nicely typed, and apparently complete, notes. Which would free you from the burden of having to write stuff down in class, thereby freeing you to listen undisturbed to the dulcet tones of the professor.

We counsel against. Knowing you'll have to take notes is one of the main things that gets you to go to the lecture in the first place. And how will you stay awake and focused if you don't have that pen to guide for 50 minutes straight? Then there's the problem that the "professional" notes aren't any easier to read through and take in than any other written document—the textbook, for example. And somehow, something always seems to be left out when you just read someone else's transcript or outline of a talk (it's the things that, in spite of yourself, you remember from hearing the lecture, things that are brought to your consciousness when you reread your own notes).

Finally, there's the real possibility of note-taking error. A Cornell professor once griped to us about how the note taker had misunderstood the professor's discussion of the four *cardinal* virtues (justice, prudence, fortitude, and temperance), instead taking down notes about the four *carnal* virtues (we leave it to you to think out what these might be!).

Review Session

Taking excellent lecture notes is one of the most important skills you can develop in any college course. In addition to furnishing you an object of study for future
tests and papers, it guarantees you an active engagement
with the lecture and it kick-starts your learning process.
Luckily, there are a number of things you can do to make
your notes an excellent rendition of the lecture, and to
make your note taking an intellectually productive, indeed enjoyable, experience. Here are our 10 best tips,
boiled down to 5 (think of it as a 2 for 1 sale!):

1. Come to class prepared to take notes, thoughtfully
and actively, for the entire period.

2. Use the professor's outlines, as well as verbal and behavioral cues, to capture the structure of the lecture.

3. Realize that things written on the board, or on Power-
Point slides, are just abbreviations and always need
explanation in your notes.

4. Get down definitions for all technical terms, and devise a system for distinguishing examples, diagrams,

and other illustrative material from the main points of the lecture.

5. Take notes for yourself—without stupid "note-taking formats"—and do it once and for all.

Producing excellent class notes is one of the easiest ways to position yourself for getting A's. And it won't take you a minute extra (beyond the time it takes to go to class). Is this a great deal or what?

CHAPTER SIX

Why Prepare? Why Attend? Why Participate?

If you're anything like most college students, you'll find yourself asking from time to time: "Why prepare for class?" "Why attend lecture?" "Why go to discussion section—and once there, why participate?" Sometimes these questions are genuine requests for information. You're short on time. You're juggling four or five courses. Not to mention a job—and a life. You want to know what the relative value is (translation: what the grade value is) of doing each of these class activities.

At other times, though, such questions are expressions of apathy, disillusionment, or exasperation. Why slave over the reading if the professor is going to present the material better in lecture—and, more to the point, without any effort on your part? Why make all the lectures if the professor tends just to summarize the reading or to do the very same problems that you've already polished off at home? And why waste your time with the discussion (or quiz or drill) sections if the TA is only going to review the stuff for a third time or answer half-brained questions from students with only one oar in the water? Excellent questions—all. To which you have the right to an answer. If not from your professor, then from us . . .

WHY PREPARE?

In most college courses you're assigned things to do outside class. Every week, without fail. Maybe reading, or problem sets, or sentences to translate. College professors sometimes call this "preparation for class," but that's really just a glorified way of talking about what you might call *homework*. The only difference between preparation and homework is that, much of the time, college instructors don't collect or grade these assignments. Hey, for most of us three or four pieces of graded material per student per semester is plenty enough.

This does not mean that there's no value in preparing. Preparation for class is not some throwaway activity or something to pass the time with while the Internet is down. It's actually the *main part* of your learning experience in any college course. Yeah, you thought that listening to the professor yakking away was where it's at. But the lectures are intended to take up less than half of your learning time. The bulk of your learning is supposed to happen as you work on your own, not just studying for tests and writing papers (which we'll be talking about in the next two parts of this book)—but also when you *prepare for class*.

★★★★☆ **4-Star Tip**

Preparation is the major part of the learning experience.

The assignments that professors give are designed to help you learn—and at a manageable and measured pace. Most professors divide up the assignments so that you have about an hour or two of preparation for each class. This sure beats having 26 hours of reading and studying before the first test;

Professors' Perspective

Most college courses have about 150 minutes of class-time instruction per week. That's about the amount of time of your average full-length movie. Think about it. We professors are charged with teaching students about American history before the Civil War or about macroeconomics—we are asked to get students to master calculus or learn to speak Spanish or figure out how to write good essays—in roughly the expanse of time equal to watching 12 or 13 movies. Even if we talk at half the speed of light, without stopping for air, there's just not enough class time to accomplish this. That's why we give assignments to do out of class. If we thought you could master formal logic without doing proofs on your own or learn about Russian literature without reading any novels, believe us we wouldn't bother you with any out-of-class assignments. If only because it would give us less to read, too. (Did you realize that we also stick ourselves with reading whenever we assign it to you?)

it's like the difference between running a mile or two every three days and running a marathon. No one ever hit the wall putting in a few hours a week preparing for class. (If more preparation is expected in the course you're taking, this would be good to know, too. Be sure to ask if you're in doubt.)

A managed pace is particularly important in courses that either have tons of reading (like English or history) or involve a gradual buildup of skills and knowledge over the course of the semester (like foreign languages and math).

The student who thinks he or she can put off reading Tolstoy's *War and Peace* or de Tocqueville's *Democracy in America* till the day before the exam is also the student who thinks he or she can eat an elephant in a single bite. And the student who leaves the memorization of the Spanish vocabulary items till the night before the midterm is either from Madrid (in which case he or she shouldn't be taking this course) or else has committed him or herself to three all-nighters—in a single night!

On the other hand, the student who conscientiously does the work as it is parceled out has a relatively manageable 50 pages of Russian literature or American history to be done before each lecture, or 10 or 15 Spanish words to be memorized before each meeting. Not so bad, when you think of it that way.

Another good thing about preparing before lecture is that it helps you understand the lectures better. Jeremy particularly remembers a Great Books lecture in which the professor tried to evaluate whether Odysseus should be considered a hero or just a two-timing (or three-timing or four-timing) louse. Who could really understand this lecture without having first done the assigned reading of Homer's *Odyssey*, having learned about Odysseus' heroic killing of the Cyclops, or his multiyear dalliances with Circe and Calypso (all while still being married to the lovely Penelope)? Equally memorable is the middle-level logic course in which the professor went over variants of the problems Jeremy had slaved over the night before. Here the professor simply *assumed* that the students had completed the assigned problems (or at least made a very good stab at doing them) and conducted the class as a communal enterprise, with each student contributing what he or she had prepared. (The prof didn't even tell the class he was planning to do this, so it all came as something of a surprise to us. At least the first time.)

SOAP *IN OUR HUMBLE OPINION . . .*

In some courses—either by choice, by requirement, or by "suggestion" from the professor—students prepare together in study groups. Is this a good idea? It can be. Studies have shown that students who work together often learn the material better than they would have, working on their own. When there are difficult physics or math problems to be solved, when there are dense primary sources to be plowed through, or when there are complicated economics concepts to get your mind around, the group approach often works wonders. Group work provides more brainpower (especially if there are "bright lights" in your group), more discipline (setting a meeting time actually gets you to study), and more peer pressure (no one wants to look like a doofus in front of his or her friends). Of course, if your study group members are not up to the task, or view studying as a kind of "social networking," you'd do better to find another group or to go it alone.

Most professors prepare their lectures on the assumption that the students have done the assigned work before the class. They have to. Whether or not it's actually true. There simply isn't enough class time to bring everyone up to speed—particularly those who haven't taken the time to prepare fully. When you haven't bothered to do your part, you're simply out of alignment with the professor: you're not on the same page, not working under the same presuppositions.

Of course, deep down, professors realize that many students in fact are not coming to class prepared. Professors recognize the blank stares, the dead silence that meets any question they ask, the student questions about even very basic points, for what they are—indications that, for many students, this is the first time they're hearing about whatever the professor is lecturing about. And then, thrown into a tizzy of worry, the professor adds extra quizzes, pop quizzes, and even test questions on the reading (including sometimes reading not even discussed in lecture)—just to ensure that the students have done their share of the heavy lifting.

If your professor adds such seemingly piddling items to the course menu, there's no need to go nuts. These things won't count all that much in the final grade, in any case. View them instead as an incentive to prepare. And as a diagnostic tool. If you bomb the reading quiz in your Classical Civilization class or the Friday vocabulary quiz in your German class, you've been given cheap feedback that you're not preparing as well as you should, or in the way that you should (remedy: you need either to work *harder* or to work *differently*). Either change your preparation methods or ask the professor or TA for tips about how to better prepare.

You Can Do It! 10 Things to Do When There's a S***load of Reading

Certain courses require many, many pages of reading. History and literature courses are notorious in this regard. Many history and literature professors think nothing of assigning reading in 100-plus-page blocks.

When you find yourself in this kind of course, try some of the following techniques to make the reading more manageable:

1. **Make peace with your lot.** Once you've picked a reading-heavy course, you're stuck with lots of reading. That's just the nature of the beast. As our favorite history professor, Lynda Coon, puts it, "Reading? That's just what historians do."

2. **Determine how much of the reading the professor really expects you to do.** Contrary to the belief of some, when the professor assigns a ton of reading, he or she often has determined that that's the proper amount of reading for that level of course. But in other courses, professors simply lard the syllabus with everything they can think of about the topic. Find out what your professor really expects: either ask someone who has taken the course or get your nerve up and ask the professor or TA yourself.

3. **Devote one to two hours to each reading session.** No one can get a meaningful amount of reading done in 15 minutes or half an hour. You need time to actively interact with the reading, watch the points develop, and let them sink in. Many times readings have a dramatic buildup, or incorporate an argument developed in steps, or put forward a main point that becomes clear only in time. Breaking your reading up into too many small pieces defeats the *construction* of what you're reading.

4. **Assemble informational clues before starting.** Often significant information is available to you, before you start your reading, about what the author is going to argue (and how he or she is going to argue it). Sometimes the professor has told you what to look for; sometimes there are "study questions" on the syllabus; sometimes there's an abstract or summary at the head of the reading; and sometimes there's a table of contents or section headings that tell you how the

article or chapter is going to be laid out. Familiarize yourself with such useful information in advance of reading the piece: clues like these help eliminate uncertainty, thereby facilitating reading.

5. Read in meaningful units. The worst kind of reading is word for word. Instead, read a whole chunk containing an idea or point—perhaps a paragraph or two in an article or textbook, or a page or two in a novel or short story. Then take a brief break to sum up in your own mind what you have read. Then proceed to do it again on the next part.

6. Strive to anticipate what comes next. The most successful readers are often able to have a pretty good idea about what comes next in a given reading. It's not that they guess or make up the following point; they just use the information already available in what has been said to form a (tentative) hypothesis about what the next point may be.

7. Look up significant words you don't know. If you encounter an important term you don't know, take the time to look it up in your notes, in the glossary of the text, or in a dictionary. But do so only if the term is strategically important, or an often repeated term in the reading. Don't break your stride looking up every little word whose meaning you're not 100 percent sure of. That's a waste of time and throws you off the track.

8. Try to maintain a decent pace. In every field there's some shared conception about the speed at which one ought to be reading the material. If you feel your pace would make the tortoise proud, find out what's reasonable or standard in the field you are studying. The professor or TA will *tell* you, if asked (after the hundreds of mandatory disclaimers that "different people read at different speeds," "there's no right or wrong speed," "you should read till you understand," etc.).

9. **Highlight or take notes, if you like, but realize these are just tools.** Many readers find it useful to highlight in colored markers, to make marginal notations, to take continuous reading notes, or to jot down outlines as they go. Good practices, all. Just be sure that devising and implementing your system doesn't become the main focus of your reading activity (we've seen students spend as much time figuring out their code for the nine colors of highlighters as actually doing the reading!).

10. **Don't panic too soon.** Reading, especially in a new field, is a skill that takes time to practice and develop. In the good case, the professor knows and has attended to this fact and has chosen the readings to escalate in quantity and difficulty. But even when it has never once crossed the prof's mind that students have to actually read the stuff, you can build up your skills by persisting and doing the reading time after time. Never say die. Improvement in reading—and diminution in anxiety—happens more often than not. We know. We see it all the time.

WHY ATTEND LECTURE?

If you've read Chapter 5 on excellent note taking, you already know that the lecture content can be a big object of study in any college course. That, in and of itself, should be enough to convince you of the value of class attendance. Not to mention all the preparation that goes to waste (or at least isn't used as fully as it could be) when you do all the reading, or complete the problem set, or memorize the verbs, but then don't bother showing up for class.

But we've found that many students need still more convincing. Especially those at the University of Minnesota, who have to brave the 25-below-zero temperatures to get to class, and their counterparts at Arizona State University, when it's 110 degrees. Somehow, to them—and perhaps to you—it seems pretty darned easy to just blow off this one class. After all, who's going to notice? (Not the professor who's busy with the lecture, and wouldn't learn the names of the students even if he could.) Who's going to care? (It's just one class and probably not that much went on, anyway.)

 Professors' Perspective

Most students have a cutting budget—no, not a cutback in their spending habits, but a maximum number of classes they feel comfortable missing. Most professors, too, have a sense of the most classes a student should miss over the course of the semester. Unfortunately there's a huge chasm between the two figures. Many students feel it would be okay to miss six, seven, or even eight classes without much adverse effect on their grade. The average faculty member thinks that two or three would be more like it. Who's right here? We'll let you be the judge. Considering that each class is about 3 percent of the lecture content . . . well, you do the math.

When thinking about attendance, it's helpful to realize that lecture classes usually fall into two main types. In the first type, each lecture covers a different topic. If you miss a lecture in this kind of course, you might hardly notice any problem, since each lecture stands on its own. Everything is fine—except for the fact that you missed a whole topic, which could appear on the exam or paper assignment. In the second type of course, the professor develops a sustained exposition or argument over a *series* of lectures. If you miss a lecture that's part of a series, you might not be able to understand any of the subsequent lectures.

Jeremy Remembers When . . .

One time I missed a lecture in my Greek Philosophy class for no really good reason (I can no longer remember if I was tired, sick, hungover, or all of the above). This was lecture number two in a series of five lectures on Plato's *Republic*. In these lectures, the professor distinguished various kinds of justice, which he designated $justice_0$, $justice_1$, through $justice_6$. It was not until the fifth lecture that I realized that missing lecture two meant that I hadn't understood a single thing in lectures three, four, and five (even though I had dutifully attended all of them). And if that wasn't bad enough, the question on the midterm was about senses of "justice" in Plato's *Republic*. That missed second lecture turned out to be the gift that kept on giving!

You may not realize it, but attending class is also extremely time-efficient. Though you might seem to be saving time by not going to class, it'll take you more than double the time to make up the material you missed. And, you probably won't learn the stuff as well as if you'd just attended class in the first place.

What to Do When You Miss Class

Even the best of students sometimes miss the lecture—and for very legitimate reasons (such as illness, emergency, car breakdown, brush with the law, etc.). If this happens to you, don't beat yourself up. Instead, follow this five-step procedure:

1. Find a student who moves his or her hand during the entire lecture (writing, we mean) and ask to borrow his or her notes.
2. Study the notes carefully, have another look at the reading, and figure out what parts of the notes (if any) you don't understand.
3. Go see the professor (or TA), make a brief excuse for missing the class (the shorter and more polite, the better), then ask specific questions about what in the notes you're not sure of.
4. Fill out your notes with what the professor is saying, because sometimes what the professor tells you one-on-one is even better than what he or she said in the lecture.
5. Resolve not to miss class again. This makeup method is very cumbersome, and the professor will get tired of doing your work for you more than once or twice.

Visitors

Visiting Professor

David Christensen, University of Vermont

Sometimes a student will come up to me and say, "Uh, Professor? I kind of had to miss class last time, and I was just wondering if there was, like, anything important I missed?" I always say (pleasantly), "Oh, no, you know we never do anything important in class." Then I just wait, and watch his face as he figures out how his question must have sounded from my perspective. He generally doesn't ask that question again.

WHY GO TO SECTION MEETING (AND, ONCE THERE, WHY PARTICIPATE)?

In many college courses, particularly introductory courses in large schools, there are section or discussion meetings held in addition to the regular lecture classes. Normally these meet one hour per week and are led, not by the lecturer, but by a graduate-student teaching assistant (once in a while, the professor teaches one section him or herself—an excellent value if you can get it). Sections are part of the college scene because they give students in megasized lecture classes a few minutes a week of small-class time—time when they might actually be able to ask questions or discuss an issue. Not to mention that sections are a great way to train graduate students to teach, without allowing them to do too much damage to the course.

Is there really a point in going to these parts of a class? Aren't they going to be a complete waste of time? The problem in answering these questions is that there's no agreement about what kind of material should be covered in section meetings. You'll find that the activities in section meeting will vary widely from school to school, and from course to course. Indeed there are often even substantial differences in what goes on in the sections of different TAs in the *same* course— either because the professor gives the TAs free rein to do whatever they want, or because the professor tells the TAs what to do and some still do whatever they want.

Nevertheless there are a variety of activities that typically go on in section meetings. Here's a list of 8 things that could go on in section—and our assessment of the value of each:

1. **Reviewing or summarizing the lecture material.** (Can be useful, especially if done in an interesting or different way. Depends on how gifted the TA is.)

2. **Taking questions** that occur to students as they look over their lecture notes. (Not so great, unless the students and the TA are very smart.)

3. **Studying new material** (for example, articles, issues, or problems) not covered in lecture. (Very valuable, especially if this material makes its way onto the test or paper assignment.)

4. **Performing analysis** that goes deeper than what was done in lecture (e.g., problem-solving, working on a new case, studying an issue in greater detail, doing a scientific experiment). (Very valuable, especially if you're honing the analytical skills that will be tested later.)

5. **Student discussion or presentations.** (Often useful, since you get to learn by doing. And usually fun, too, since who doesn't like to hear him or herself talk? Could also help your class participation grade and maybe even your test grade, since some professors, including us, think that discussion content is fair game for tests.)

6. Working on group projects. (Can be good, depending on the group, the project, and how social an animal you are. Always important for the grade, for who would devote class time to projects and then not grade them?)

7. Counseling and practice for tests. (Now you're talking. Real value. Definitely not to be missed. TAs are a prime source for extra hints. They are much less battle-hardened than your average professor [and usually closer in age to the students], so they often say a lot more than they should.)

8. Administration of quizzes and tests. (The mother of all section activities. Miss this section and you're fried.)

But sometimes, in spite of the professor's (and TA's) best intentions, the section meetings seem to be a total waste of time. You show up and the TA just goes over the same old stuff you've already heard in class (and weren't all that interested in the first time). Or the whole time in section is devoted to students asking questions that are either totally stupid or completely irrelevant to the course. Or the TA is a complete idiot and you could do a better job running the section in your sleep. Surely in *these cases* the best course is to skip section and hit the bars—sorry, we meant the books. Isn't it?

Turns out not. The first thing to do is to try another section or even a *few* other sections. There's wide variability in quality among TAs, so you should shop: there's usually a big selection in most intro courses. Just don't TA-hop much beyond the first week or two of class. Get settled into one section, and then don't be stepping out on your TA by visiting other sections. Some students think that attending multiple sections will give them an edge, but this is a case where more isn't more. More is more confusion. One meeting a week with a good TA should be more than enough.

Don't give up on sections, because most often your section leader is going to be your *grader*. And one of the most important ways to get good grades is to get to know your grader. Because—as we'll be explaining in upcoming chapters—the person who can best help you get good grades is your very own grader. But how much is your grader going to want to help you if he or she doesn't know you from Adam (or Eve)? So haul your butt to section regularly, whether it's a fantastic learning experience (which it could well be, especially if you take the time to scout out a good TA) or just a time to contemplate your navel (in which case you still get to know, and court, your TA, who is also your grade-giver).

And once you're already going, it would be a bang-up idea to participate. To answer the questions that the TA (or professor) might be asking. To throw in your two cents worth from time to time and to join the community of scholars (or at least fellow students yapping in the discussion). To think, and to talk, and to think and to talk some more. Hey, this is a *discussion* section and you've got a mouth!

When you participate you learn better. Just listening to someone else talk is never as good for learning as talking yourself—just as watching someone else drive won't help you learn to drive as well as getting behind the wheel. Besides, participating is more fun. Why sit on the side like a wallflower when you could be out there dancing, or at least talking?

Keep in mind that participating in class can have an impact on your grade. That's because many classes devote a portion of the grade to "class participation." When instructors calculate the final grade, around 10 to 20 percent of the students are on the borderline between two grades. Professors (and TAs) often use class participation, either officially or unofficially, as a way of deciding what to do in borderline cases, say at the interface between B+ and A−. And at schools that don't use pluses or minuses (which is most colleges), this translates into deciding whether to give a student a B or an A. If an instructor can recognize a student's face—if the TA knows that you came, and saw, and (even if you didn't succeed) at least tried to conquer, he or she is likely to give you the benefit of the doubt and put you on the greener side of the grade fence. Especially if you follow our . . .

Top 5 Do's and Don't's for Section Participation

DO'S

1. *Do* stay on topic. Raising points from left field is not to your advantage.
2. *Do* talk less if you find yourself talking all the time, or if you hear people groaning whenever you open your mouth.
3. *Do* feel free to offer a follow-up comment if it seems appropriate. There's no rule of one comment per person.
4. *Do* prepare for class discussions so that you can offer knowledgeable participation. It's not only about quantity, stupid, it's also about quality.
5. *Do* politely apologize and pass if you are asked a question you haven't prepared. It's generally not a good idea to try to B.S. your way through.

DON'T'S

1. *Don't* view class participation as a competitive sport. The point here is to share your ideas, not beat down your classmates.
2. *Don't* ask questions just for the sake of asking questions. You don't have to "mark your territory" every time you show up for section.
3. *Don't* announce in class that you haven't prepared (if preparation is expected). No one wants to hear your two-bit excuses.
4. *Don't* be shy if you have prepared for the class. You'll be helping out if you make a contribution.
5. *Don't* stress out if you make a mistake—it's not that big a deal. Besides, saying wrong things is participating, too, and most section leaders will recognize a sincere effort when they see one.

One final tip. Once a week you should take stock of what you've done the previous week in each of your classes. And what you will be doing in the week to come, and how it relates to what you've just done. No, we don't mean redoing the reading, copying over your notes, listening to tapes of the lectures, or going over the material with your study group. Nothing nearly as elaborate as that. All we're suggesting is that each week you spend a few minutes locating where you are in the course, then regaining your bearings within the structure of the course. You might review the titles of the lectures in your notes. You might glance at the syllabus to see what lectures are scheduled for next week. You can do a reality check to see if there are any pieces you missed in the past week. And you might check to see if there are any papers or tests coming up in the next week that you ought to be starting to think about.

The process of taking stock is not something you'll find on any syllabus. No professor will tell you to do this. But most successful students do this without even being aware that they're doing so. And now you know to do it, too.

Review Session

Though it might seem otherwise (given the prominence it assumes in many college courses), the lecture is only one of a series of course activities that take place each week. All of which your professor is thinking you will do, at least if you are to excel in the course.

As a rule of thumb, professors expect an hour or two of serious preparation for each hour of lecture or section. These preparation activities are valuable for two main reasons: they break the subject matter of the course into bite-sized pieces, and they give you a leg up on the lecture by familiarizing you in advance with the concepts and points to be treated there. Unbeknownst to many students, professors also expect that you attend the lion's share of the lectures. Whether each lecture treats its own topic, or plays a role in an ongoing series of lectures, no professor thinks it's okay to blow off more than a (small) handful of meetings. And often, professors have planned some specific activities or curriculum for the *sections* (when there are sections). This means that you're expected to show up for (and once there, participate in) almost all of these, too.

The answer to the questions "Why prepare?" "Why attend?" and "Why participate?" is that a college course is designed as a series of mutually reinforcing activities, all of which need to be done if you are to learn well. And as for the apathy, disillusionment, and exasperation sometimes experienced at the big U—these will evaporate quickly when the work you did in preparing for class, attending all the lectures, and going to (and participating in) the sections results in that shining A on your first exam or paper.

PART 3

THE EXAM

CHAPTER SEVEN

13 Best Ideas for A+ Test-Preparation

The professor announces that the first test will be held a week from Thursday. Or maybe you knew it all along from the syllabus, but had put it out of your mind until now. Not to worry, you know just what to do. You've been taking tests since you were six years old, and you have your routine down pat. Read the textbook again, make up flash cards, charts, outlines. You could do it in your sleep. But wait. Before you shift into automatic pilot and study in the way you always have, think again.

College tests are a breed apart: there could be essay questions, you could be asked questions about the readings and the discussion sections, not just the lecture or textbook, and, most important of all, the professor might be expecting that you actually understand the material, not just parrot it back. Yup, it may be time to update—and upgrade—your preparation techniques. We've got a baker's dozen of ideas about how to make the leap. Guaranteed to work, or your C's cheerfully refunded.

IDEA #1: Spread It Out

Many students assume that the closer to the test they study, and the more hours they put in at a sitting, the better they'll remember the material they need to know for the test. That's why in most large classes the professor will spy a cadre of students who have stayed up all night and keep studying until the professor has to literally rip the notes out of their sweaty hands to get the exam started. This is actually the worst possible way to prepare for a test. You don't learn well and it's an unpleasant experience to boot. Both for you and your classmates, who have to put up with you showing up for the exam looking, and perhaps even smelling, like the back end of a bus.

What's really best for learning is to spread out your studying over a larger number of smaller sessions. We suggest starting your studying (when possible) about a week before the test. Look over what you need to know, then divide up the material over three to five study sessions spaced over the week. The sessions needn't be super long (about an hour or two each is generally sufficient)—just enough to cover the portion of the material you need to cover. Each session (except the first) should allow some time for a brief review of the material you've already studied. But the bulk of each session should be spent on studying some new material—always forging ahead to that ever-looming test.

IDEA #2: Triage Your Time

It's the easiest thing in the world to find yourself up against three or four midterms in the same week. Which is why you need to *triage*—to sort and allocate your study time based on what sort of grade you're getting in each course (if you already know), and on where you think your study resources can be most profitably deployed. Everyone needs to do this. No one (no matter how smart) has infinite time to study.

Some common time-allocation strategies are clearly deficient. At least if you want to maximize your chances of getting the best possible grades. The "first come/first served" plan—where you study as hard as you can for the first test of the week, then when it's over start on the second exam—runs into trouble when the tests aren't evenly spaced or don't demand equal preparation. The "I'll start with the easiest one" (or its friend, "I'll start with the one I like best") isn't a winner either, since you might never get to the more challenging or less enjoyable course (and in any case it's often best to start with the really hard or unpleasant task—ask any runner or dieter). The "I'll spend all the time on the most important course" and its brother, "I'll spend all the time on the course in my major," could be fatal to your overall grade point, if you ace that "most important" course but bomb the other four. Finally, the "I'll devote all my time to the course I'm doing *worst* in" can blow up in your face, if you manage to significantly improve your grade in *that* GPA-lowering course, but only at the expense of lower grades in the courses you had been doing better at.

The best idea is to apportion your time based on a balancing of various factors:

- ☑ Determine how much preparation time each course really requires. There's no point in budgeting a large number of study hours to a course in which there's really not that much to prepare, or for which (who knows how) you've already prepared to the hilt.
- ☑ Plan to devote a significant amount of preparation time to courses that are especially *important* to you—either

because that's what you're most interested in studying; because the course is required for, or an important part of, a major (or minor); or because the course teaches skills that are important for you to know (for your other courses or just in life).

☑ Devote at least some study time to required courses or (gasp!) courses that for one reason or another you're not enjoying. These, too, can make up one-fourth or one-fifth of your GPA for that semester.

Above all, think it out. Make a plan. Balance it off.

IDEA #3: Scope Out the Scope

The single most important thing you need to know before starting to study for the test is what exactly is going to *be* on this test. For example, what classes or topics are to be covered? Will the test include material up through the class immediately before the exam, or will some lectures be held over till the next exam? Will the test include material from the readings or only from the lectures? What about the content of the section, discussion meetings, or labs—are these fair game, too? Finally, for a test that is not the first one in the course, will it be *cumulative* (that is, cover all material back to the first day), or will it be *sequential* (that is, take up only the content since the last test)?

Most professors give lots of information about the scope of their tests. They have to, if only to forestall complaints that the test was unfair because the students weren't told in advance what was going to be on that test. So it shouldn't be that hard to scope out the scope of the test. Usually you can find the information on the syllabus and/or in the class in which the professor explains what's going to be covered on the test. (Sometimes this is done by the TA in the section meeting, so here's an extra-special reason to make discussion section, especially before a test.)

Be sure to take copious notes on anything the instructor says—or mumbles, muses, or offers as an aside—about the test. If ever there was a moment for you to be a prizewinning note taker, this is it. All this might seem pretty obvious, but in our experience large numbers of students don't bother with this. Also, when at 10 P.M. the night before the test you can't remember whether the professor said that a certain issue would or would not be on the test, simple turns of phrase captured in your notes (not to mention explicit assertions) can take on enormous importance. (If the professor didn't bother to tell you what the test would cover—out of laziness, inadvertently, or on the cockamamie theory that "part of studying for the test is *figuring out* what will be on the test"—then he or she deserves all the pestering you and your classmates can offer. Later we'll tell you a number of "persuasion techniques"— none of them violent—that you can use to get the professor to "give it up.")

IDEA #4: Get Your Hands on the Treasure

Though the scope of the test is a big thing, the real prize in the test-preparation business is getting your hands on an actual copy of the exam. Now don't get us wrong. We're not suggesting you break into the office of the professor and try to steal an advance copy of the exam. That could get you thrown out of school, if not into the slammer. But there are a number of legal ways to get copies of previous exams, and these can be of tremendous benefit for your grade.

Sometimes (nice) professors will give out samples of their old exams. Sometimes old exams will be kept on file at the library or in the files of fraternities, sororities, or dorms. Sometimes your friends will have taken the course in previous years and will have kept their copy of the exam, or at least will remember the test structure and its questions. This is definitely the time to go file-digging or friend-making.

✔⁺ *EXTRA POINTER*

> If you're using an old exam to help with your preparation, just be sure the exam is from the same course and, more important, from the same professor in that course. An old exam from a different professor is usually totally useless, since professors have considerable discretion in setting up the test format and the types of questions for even the same course.

It's also very common to have professors who give out *study questions* or *study guides* in advance of a test (the difference being that study *questions* are lists of sample questions for study, while study *guides* mark off areas or topics for study but do not furnish any questions). These "previews" of the test are an enormous gift (even though fewer than half the students realize how big a gift they are). Very often the actual exam questions will be selected from (or at least closely modeled on) the samples given. And even when you're not *that* lucky, the study guides and study questions can at least narrow the terrain of what's going to be asked, and hence, of what you need to study.

Professors' Perspective

Many professors are too overworked to make up new tests each year, especially tests that are very labor-intensive to construct, such as complicated multiple-choice questions, detailed problem sets, and essay questions with lots of parts. And in many fields it's hard to think up "good questions"—questions on the right level of difficulty, suitably central, doable in the requisite time, and designed to distinguish different levels of ability—year after year, course after course. So the professor has to either recycle questions from his or her stable of questions, or make up questions that are minor variants of the questions on the sample exam or study guide. Net result? The students who use this antecedently available information to structure their preparation and construct test questions of their own are at an incredible advantage relative to the students who either haven't even gotten the study sheet or have viewed it as a mere "announcement" of the test to come.

IDEA #5: Line Up Your Ducks

This one's simple. Before beginning your first study session, gather together all the materials you'll need to study. Your lecture notes, section notes, reading notes or outlines, any relevant books or articles, problem sets, homeworks, quizzes—and of course previous or sample exams, study questions, or study guides. All of these need to be complete and in order. How can you do well on your U.S. History test if your lecture notes on the Spanish-American War come before your notes on the Civil War? Or if your notes on that article about the significance of the Alamo have found a home on your laptop in the middle of your class notes on the Constitutional Convention? (If you have no idea what the proper order of these events is, don't worry. Your school has an American history requirement in which you'll learn this sort of stuff.)

Being scattered makes it harder to study. Not only do you waste lots of time looking for the pieces you need, your lack of organization can prevent you from seeing all the data and the relations among the data. In the worst case, you could even miss the whole plot of the course, or of some section of the course. Often professors are looking to see that you understand the *development* of the ideas in the course, and the *relations* among the different parts of the course (or among the points in a single part). Which you can't do when your notes are a pig mess. Here's where all that classwork, all those times you walked the five miles to class in the pouring rain and took notes till your fingers turned numb will have their payoff. You have the data. Don't jumble it away.

IDEA # **6:** Review, Don't Redo

Throughout your preparation activities you should have one overriding thought: *keep looking back*. No, we don't mean forget about the upcoming test (how could *that* be a useful idea?). We mean: *Review, don't redo. Go over, don't do over.* So don't do the reading again. Don't recopy your notes. Don't generate notes on your notes. And don't be doing stuff from scratch. This isn't the time to be doing the work for the first time. It isn't the time to do extra reading or more research (unless specified by the professor).

Preparing for tests isn't at all the same thing as preparing for class. When you study for class, you are learning facts and concepts for the first time. But when you study for a test, you are relearning, synthesizing, and assimilating what you have already learned. If you find yourself redoing stuff, or learning new stuff, you aren't studying in an efficient manner (or in the way the professor intended). And you aren't putting yourself in the best position to get an A on that soon-to-be-coming exam.

IDEA # 7: Don't Be a Voyeur

Lots of students equate studying for exams with simply rereading their notes. They sit. They read. They absorb. That's all. And that's all well and good. But it's not enough. Because studying for exams isn't just about looking at and internalizing your notes. It's about *doing* something—actively processing the information that is in your notes. When you're studying, don't be passive, be active. Don't be a voyeur, be a participant.

Begin by trying to locate the main points of the course—that is, those points that *your professor* thought were the main points in the course. As opposed to high school, in which your teacher might simply have synthesized or summarized various points, in college your professor probably has forged his or her own take on the material. What to leave in, what to take out. He or she might have even (gasp!) done research on the material. And have a point of view.

Usually the main points are the ones the professor spent a good deal of time explaining, discussing, and probing. The ones he or she kept going on and on about. Now, this might come as a shock to you, but normally professors test on what they consider the most important points in the course. Think about it for a moment. If you were trying to test in order to see if someone had learned what you taught, would you (a) ask him or her questions about what's most important, or (b) forget about what's most important, and ask questions about minute details that were touched on for less than two seconds? Look, it's not that we couldn't make a test so tricky that everyone would get every question wrong. It's just that we're actually trying to *teach*.

★★★★☆ **4-Star Tip**

Unless told otherwise, study the main points of the course. Most professors focus their exams on the most central points.

As you read over your notes, get the wheels and springs of your mind going. Be inquisitive. And deep. Ask yourself questions like: What was the main point of this lecture and how did this lecture fit in with other lectures in this part of the course? What were the key concepts emphasized, and the methods used, in the various lectures? Did the reading (and/or section or lab) reinforce the key ideas of the lecture, or did it introduce new material? Think beyond the surface. Focus not just on facts, but on *relations* among points. And pay attention not just to the points made, but to the *methodology* (of that field) used to generate those points.

IDEA #8: Make Sure Your Answer Is in the Form of a Question

Most students' biggest fear as they prep for an exam is that they won't know the answers. It shouldn't be. What they should be really terrified of is not knowing the *questions*.

One of the most important parts of being a successful (that

is, an A-getting) student is being able to *anticipate* what the professor is going to ask on the exam. Once you can do this, you can prepare much more effectively. You can focus your study on—and as we'll see later, actually try out—the kinds of questions that will be on the test. You can avoid information overload and marshal your limited resources to studying what's really important. And you can avoid stress, lots of stress, because you're driving the cart, not being pulled by it.

But how do you discover what evil lurks in the recesses of the professor's mind? Well, every course is littered with clues and foreshadowings of what the professor is going to ask on the exam. Often the information is right out there in the open. If the professor has given out a sample exam or study questions, mine these for all they're worth.

Pay particular attention to the *structure* of the sample exam and questions: Are there multiple-choice and short-answer questions in addition to essays or longer problems? What percentage of the grade is each worth? Then consider the *type* of activity tested in each sort of question: Is mere memorization required, or is there some further expansion or processing needed even in the short answer questions? For essays, what sorts of tasks are being asked? *Compare* is different from *contrast*; *describe* is different from *evaluate* (which in turn is different from *analyze*); *advance a hypothesis* is different from *summarize*, which also is different from *state*. Professors go to great lengths to pick the exact verbs in these sorts of questions, so make sure you are 100 percent sure what to do with the samples. (You'd be surprised how often professors are forced to give C's simply because the student didn't do the task[s] asked.)

Finally, try to figure out whether the professor is testing by *sampling*—that is, by selecting a few representative issues or questions, then assuming that the way a student answers these questions is indicative of how well he or she understands all the material; or whether he or she is testing by *broad coverage*—that is, by attempting to test in some way all the different topics or issues that have been covered in that part of the course. Depending on how you size up the situation, you may choose to study all the material in a somewhat more superficial way, or to hone your skills by working only on a smaller number of representative questions or problems.

But sometimes you're not so lucky. In some courses the professor just doesn't give out any handouts to help you prepare for the exam. And the TA is no better. For some reason (who knows what) the instructional staff thinks either it's just obvious what will be on the test, or if it's not, the students will somehow just figure out on their own what they need to study.

In this case, what you need to do is scour all available sources in the hopes of getting some glimpse at what might be on the exam. Maybe the syllabus or the lecture titles offer hints. Maybe one day your professor was lecturing and stopped to muse, "You know what, that would make a good question for an exam" (*translation:* like the exam I'm making up tonight). Or maybe the TA (who often has a role in preparing the test or an advance peek at it) suggests, "It might be a good idea to pay some attention to _____ in your studying." Instructors see this as a "nudge, nudge, wink, wink" moment. As a major hint. Students tend to assume that no instructor would really be dropping such an obvious hint and hence pay no attention to what's said. Pity. Just another teacher/student miscommunication moment.

✓+ *EXTRA POINTER*

The first test in a course provides a sample for *subsequent* tests. Students who throw completed tests in the trash are throwing out one of their best resources for striking gold on their next test.

What the Psychologists Have to Tell Us about Learning and Memory

Whenever you start a project, it's good to know the tools you're going to be using. So it makes sense that as you start to study it would be helpful to know a little bit about how human learning and memory work. Luckily there are folks who know this sort of thing. Like our psychology friend,

Visitors

Visiting Professor
Diana Nagel, Northwest Arkansas Community College

Learning is an active, constructive process, in which you constantly add new content to what you already know in an attempt to build a meaningful whole. Using proven techniques that reflect our human psychological reality can make your study time more efficient and productive, making it easier to get good grades with less work. Here are five ideas:

1. Active learning. People learn more, and learn better, when they are actively and dynamically exploring a topic than when they just sit back and soak in the passing show. When studying, constantly ask questions about, and forge relations between, parts of the material you are studying. Probe and explain, think out and test.

2. Elaborative retrieval. People remember better, and for a longer period of time, when they structure information to form meaningful wholes (rather than simply repeating the ideas again and again). That's why it's easier to remember your phone number when you group the numbers into units of three, two, and two than when you just say the seven digits one at a time, over and over. When studying, try to relate what you're studying to things you already know in the course, making a careful mental note of what's new in the new content. Always provide some semantic organization—that is, organization based on meaning—as you study; never just repeat and repeat.

3. **Diversity of retrieval cues.** People remember best when they use different sense modalities (seeing, hearing, etc.) as they study. By using multiple senses people process material more fully and create more ways to retrieve the content from memory. Move around; talk out loud; make up songs and mnemonic devices; imagine the smells, tastes, and sights of the material. Make it all as vivid as you can, and use as many senses as you can.

4. **State-dependent learning.** People remember material best in the place they first learned it. If possible, it's best to study in the same place you first learned the material—the lecture hall of the class—and in which you're going to have to take the test. If you can't study there, at least study in a series of *different* places, so your memories won't be correlated with a single place that is not the test room.

5. **Memory consolidation.** People collect and strengthen their memories as they sleep, literally fortifying the connections among the neural path ways in our brain. Space out your study sessions, with time for breaks and sleep in between. Cram right before an exam only if absolutely necessary. And if you must cram

(and won't be able to benefit from any memory consolidation), focus on the most important stuff. You might at least be able to remember *that*.

Students' study techniques vary. Figure out what's best for you. But don't fight your mental hardware.

IDEA #9: Keep Things in Proportion

Many times tests have different parts, each with different tasks you have to do. These parts generally have different point values or percentages for the overall grade on the test. You'll have the best shot at top grades if you apportion the time you spend preparing the various tasks according to how much each counts for the grade. For example, Lynn typically gives tests with both a slide identification and an essay section; the slides are worth 20 percent of the grade and the essays 80 percent. Given this fact (and Lynn is totally open about it), anyone with half a brain would spend about 20 percent of his or her study time going over the slides, and about 80 percent preparing sample essay questions. But Lynn routinely encounters students who obsess over the slide memorization to the point of almost completely neglecting the essay questions. Not such a swift idea. Jeremy, on the other hand, likes to give tests that are evenly weighted between tracing a theme through three philosophers, and analyzing a single argument within a text. But some students end up studying for only one of the two tasks—generally the one they think is easiest for them. Which doesn't usually have such a good end.

So check out the sample exam (or study questions or study guide) in *your* course. Figure out how much each kind of question is going to count—and then apportion your study time accordingly.

IDEA # 10: Don't Confuse Wheel-Spinning with Preparing

Some students like to prepare a page or two of notes, or a brief outline or chart, covering all the materials they need to master on a test. That's all well and good. But some students have test-preparation methods that veer into the obsessional. They spend hours organizing. They outline their outlines. They highlight (then rehighlight) their books and notes with more colors than are in the 96 box of Crayola crayons. They make incredibly elaborate lists and "clustering diagrams." By the time they've finished all this, there's almost no time left for actual studying. *Make sure your test-preparation tactics all advance your learning of the material.* And are not just disguised ways of putting off the real work. Remember, you're the easiest person in the world for you to deceive.

IDEA # 11: Work with the Ones Who Know

As you sweat your way through your test-preparation, don't forget that the professor and TAs in the course are willing and able to help. They have even set aside time to do this. And maybe even set up review sessions.

COLLEGE SPEAK

A review session is a meeting, held by a professor or TA before a test, to help the students prepare for that test. Review sessions usually take place outside of class, often in the evening. The format varies, but they typically include brief overviews of the content presented by the instructor, the instructor's responses to questions from the students, and/or the instructor's doing sample questions or problems.

Going to the review session is always worthwhile. We've seen cases in which a student has asked the very question given on the test and the professor has dutifully answered it. We've seen cases in which the professor has given out study questions, told the students he would not answer the study questions in the review session, and then still gone ahead and answered them. That's why you should be there, taking notes—yes, doing the whole number. Simple rule of thumb: 1 hour of review session equals 97 hours of preparation on your own.

Then there are *office hours*. Professors (and TAs) typically set aside extra time to deal with increased student flow around exam times, so they'll be expecting you. Do not feel that you are imposing—professors like human contact, and in any case, they're being paid for these consultations.

EXTRA POINTERS

You might wonder what exactly to say during an individual meeting with the professor before the test. Well, if your professor hasn't graced you with study questions or a study guide, you might show your professor a

sample question you have prepared and ask straight
out, "Is this the type of question that conceivably could
come up on the test?" Listen very carefully to the professor's answer, paying special attention to nuances
such as "Well, *that* question would probably never
make it onto the test. . . ." Feel free to (very politely)
ask a follow-up, such as "What sort of question *might*
find itself on the test?" Sometimes after seeing you
make a good stab, the professor will cough up a perfect
rendition of the type of question on his or her tests.

If there are study questions, it would be good to ask
questions about one or two that you are having trouble
understanding. Here, asking a more specific question is
often better than just asking the professor to answer
the study question. Usually professors won't directly
answer study questions, but once you've gotten the
prof to start talking about one aspect of the question,
he or she will usually keep going till the whole thing
has been covered. (One thing that's generally not such
a great idea is to come in to an office hour with
written-out answers to the study questions and ask the
professor to read your answers. Professors want to help
you with the process of learning, and, in any case,
don't want to start grading before the test even starts.)

IDEA # 12: Load Up Your Head

It's not enough to study. Even if you study actively. At a certain point you need to load up your head: take the funnel,
stick it in your ear, and pour the data into your cranium. In
other words, for certain tests you have to *memorize*. In some
courses, lots and lots of stuff. So you can spit it back on the
test.

Everyone has his or her own techniques for memorizing. Some people say things out loud; some make up acronyms; some correlate key information with some weird mental image. Whatever floats your boat, use it. But be aware that these techniques work only when genuine memorization is needed on a test. Some college tests involve very limited memorization and stress higher-level thinking skills. Some make memorization only a small part of the overall point allocation.

IDEA #13: Take a Test-Drive

We've left our best idea for last. You wouldn't buy a car without taking it out for a test-drive, would you? So why would you take an exam without doing a trial run? After assembling your materials, poring over them for the better part of a week, studying actively, and trying to think up questions, prepare to take an actual exam. At home, before the real test.

If you're lucky enough to have an old or sample exam, take this practice test under test conditions and timing. Seal the doors, turn up the heat, set the timer, and work away. Even if you don't have the genuine item, pretend you do. Take some questions from the study questions, or problems from the homework, or questions from the textbook, or questions you anticipate might be on the test, and construct your own sample exam. Be your own professor. Make up your own test. Then take it.

It's very important that you write it out. And that you spend the time you'll be allotted at the exam. Not more. It might seem stupid, but simulating the test experience is the single most important preparation activity that students fail to do.

You can learn all sorts of things from taking a practice test. Lots of unexpected things can happen. Maybe you didn't have enough time. Maybe you spent too much time on one thing and didn't have enough time for the other part. Maybe you had no idea how to answer the question or parts of the question (even though you concocted the question yourself). Maybe you couldn't remember the examples or other subpoints that needed to go into your answer. Or maybe you panicked when some question was harder than the others, or when you simply couldn't remember what you thought you would.

Whatever the case, if you mess up, don't feel bad. This is a good thing. Now you know what difficulties you might have in taking the test. And while you still have plenty of time to fix things up—to take corrective action so that nothing like that happens in the real test. The test that counts for the grade.

Pop Quiz
[aka Review Session]

Can you believe it? Now? A pop quiz?

Match the snarky version of each idea for A+ test-preparation with its true meaning.

1. Take a test-drive.
2. Load up your head.
3. Work with the ones who know.
4. Don't confuse wheel-spinning with preparing.
5. Keep things in proportion.
6. Make sure your answer is in the form of a question.
7. Don't be a voyeur.
8. Review, don't redo.
9. Line up your ducks.
10. Get your hands on the treasure.
11. Scope out the scope.
12. Triage your time.
13. Spread it out.

a. Divide your study time into smaller sessions over the course of a week.
b. Balance your study time over your various courses.
c. Find out exactly what's going to be covered on the exam.
d. Get the old exam, or the study questions or study guide.
e. Assemble your materials before starting to study.
f. Don't do anything for the first time or from scratch.
g. Study actively.
h. Use available materials to construct possible test questions.
i. Divide your study time for the test according to the points for each part of that test.
j. Don't waste your time on needless organizational activities.
k. Go to the review session or an office hour.
l. Be sure to memorize what needs to be memorized.
m. Write out a practice test.

Answer key:

1 = m, 2 = l, 3 = k . . . (you get the pattern)

CHAPTER EIGHT

Acing Exams by Adjusting Your Attitudes

Think you know all there is to know about test taking? That you learned all you really need to know in fourth grade? That the secrets to acing exams can be summed up in five tips: • Pay attention to qualifiers (like all, always, most, identical) • Notice negatives (such as no, not, none, and the prefixes il- and im-) • Choose the best response (especially among a number of good responses) • Eliminate answer options that are not grammatical • Mark only "sure things" first, then make three "passes" through the test?

Well, you're not 100 percent wrong. Tips like these—taken from an actual college guide to multiple-choice questions—are broadly true. The trouble is, they're unbelievably simplistic. And everyone in college already knows them (and has known them for about 10 years). These, and similar techniques (which you often hear bandied about near test time, or in "study aid" books, and sometimes in "learning resource centers"), are simply not going to help you ace college tests. No "techniques" will. What you need is an attitude adjustment. A mind-set makeover. An emotion overhaul. Out with the old, in with the new!

OLD ATTITUDE: "This is going to be like the Spanish Inquisition."
NEW ATTITUDE: "A working session? How nice . . ."

Do you find test taking very stressful? Of course you do. And why wouldn't you? You've accumulated so many bad memories from years and years of test taking that probably just hearing the word "test" is enough to send your body into "fight-or-flight" mode. It's perfectly normal. It's perfectly natural. But it's not the best idea to go into a test in a state of terror. It's not good to see tests as times when you get pounded with question after question until you collapse in a puddle of blood, sweat, and tears and are carried off to the stake. That kind of attitude would make anyone incapable of doing the clear, collected, and rational thinking that nets top grades on the test.

So stop thinking of tests as modern-day inquisitions. Start thinking of them instead as *working sessions*—opportunities to work on problems or interpretive issues that have come up in the course. Times when you get to pull together what you've been taught up till now, or when you can show off what you know. You've had working sessions before, either when studying on your own, or working with your study group, or consulting with your professor during an office hour. The only difference this time is that you're taking your working session into the torture chamber, um, we mean classroom.

Sure, we know that this time you're going to be graded, it's going to be more rushed, there's no one to help you, and you can't just peek at the book when you don't know what to say. But fundamentally, the activity is the same. It's working out in a coherent and forceful way your answers to a series of questions. Questions that (with any luck) you've prepared and are fully equipped to handle.

Don't psych yourself out even before you start. Turn the volume down. Even a small change in attitude can make an enormous difference.

But how do you make this attitude adjustment? First thing is to come into the test room dressed comfortably, and with time to spare. Stretch a little bit; then find a comfortable place to sit—no, not that seat in the middle of the middle (you're no sardine), but that nice seat in the front with the good legroom. Take a few deep breaths, get out your materials, then get your mind around the task at hand. Focus. Prepare yourself to work. Realize that you will feel a buildup of emotion and, perhaps, stress as you work through the exam questions. That's just the nature of the beast. Once you are aware of this and prepare for it emotionally, you'll be able to remain in control for the whole exam. Rather than panic and throw out cartloads of points.

Above all, bring a beverage. (We surely hope you are not taking a class in a fascistic environment that prohibits food and drink in the exam room.) Everybody needs a beverage at a working session. It establishes a basic level of comfort. It defuses the tension. It can keep you alert (if you've chosen the nectar of the gods, freshly brewed coffee). Or nourished (think protein smoothie). Or comforted (think McDonald's milk shake). That beverage should stand as a sign to you that you're there for a work session, not a torture session. Who ever heard of serving drinks at an inquisition?

OLD ATTITUDE: "Hope I don't screw this up."
NEW ATTITUDE: "Tests? My forte."

Some students come into the test with a defeatist attitude. "I'm no good at tests," they think. "I don't test well. I never do well in this subject. I never do well on essay tests. I never do well on multiple-choice tests. I never do well on short-answer tests—and come to think of it, I'm not so hot on fill-in-the-blanks either . . . and then there's matching up the columns—I always mess up on those ones." You get the picture.

This kind of pity party can be fun when you're with friends or after a disastrous date, but in the test-taking context negative thinking easily becomes a self-fulfilling prophecy. You *think* it's true, so you *make* it true.

If you encounter a somewhat harder question—which you're bound to do in anything but the easiest courses—those negative thoughts spin out of control. You start to think, "You know, I was right, I'm really not that good at this. I'm going to screw up this test. I'm going to get a bad grade. Which will screw up my GPA. You know, I shouldn't have gone to college after all." You hit one bump in the road, and pretty soon you're not only dumping the exam, you're bagging college as well.

Break the chain. Even a small change in those self-defeating (and probably false) attitudes can make an enormous difference in your results. Realize that this test is something you *can* do. There are plenty of people in the room, right next to you, who aren't half as well prepared as you, who aren't trying nearly as hard as you, and who aren't even as smart as you. You might not know it, but trust us, it's true. Keep in mind also that the exam is designed to test what the professor thinks the students should know—and *are capable* of knowing. Especially if you've made the classes, taken excellent notes, mastered the reading, and done A+ test-preparation—in short, if you've taken in even half of what this book has suggested so far. So buck up! And leave the self-criticism, self-doubt, and self-flagellation to the boneheads around you.

OLD ATTITUDE: "Aha! Just what I predicted."
NEW ATTITUDE: "Aha! Just what I predicted
            (still, you can't be too careful)."

Sometimes you open an exam and it's déjà vu all over again (apologies to Yogi Berra). You knew all along the professor was going to ask this question. Or this test is, word for word, the sample test or last year's version. This is great. A slam dunk. You spring into action. You can answer these questions with your arms tied behind your back. You could do this test blindfolded.

But hold on a second. Take some time to sip your beverage (you did bring a beverage, didn't you?) and carefully check over the questions. To be sure that the test really *is* what you predicted it would be. Because sometimes, even though the test looks familiar, there are some changes. Really important changes. But you miss them because you see only what you're expecting. You were expecting the professor to ask you to *compare* two items, so you don't notice that this time you need to *contrast* them. You were expecting to write on three poems, so you don't notice that the question asks about four. You were expecting to prove "x," so you don't see that on this test you have to prove "*not* x." You're so sure that the professor is going to ask you about some topic, you automatically assume that the question is about *that*—even when it's not.

So adjust your attitude and look before you lunge. Read the exam question, and all its parts, carefully before beginning to work. If the question is really exactly as you predicted, go ahead and dive in. Enjoy the fruits of successful anticipation (or guesswork). And if your predictions were only slightly off, pat yourself on the back for being close, then get down to dealing with what you actually ended up with.

OLD ATTITUDE: "I can't believe he (or she) asked
                  that."
NEW ATTITUDE: "Something new? I'll grab the bull
                  by the horns."

There's nothing worse than the shock of getting the exam paper and seeing questions you never imagined seeing. You sit there frozen. Like a deer caught in the headlights. Yes, you've prepared and done your best to anticipate the questions the professor would ask. Or maybe you didn't. Whatever the case, you feel like you've just been punched in the stomach. Hard.

We feel your pain. We share your shock. But take it easy. Take a couple of deep breaths (a little extra oxygen never hurt anyone). Take some time to think. Look over that question again. Are you sure you understand exactly what's being asked? You might just be overestimating the complexity of the question. Sometimes in the heat of the exam students put a weird spin on what the professor intended to be a very straightforward question. At other times the professor is asking a question in a slightly different way than you were expecting. Often, if you think about it, it turns out that you actually knew the answer all along. Another possibility is that the question asked, while genuinely new, is an *application* of something you've already learned or some skill you've already developed. No need for panic here, since you have the tools to confront the new case—if only you realize it to be such.

Lynn Remembers When . . .

One time I was taking a course on Baroque Art History. A good friend told me that the professor always asked a question about the very obscure painter Bartholomeus Spranger. Figuring I really had a leg up this time, I joyfully approached the test, only to have my hopes dashed when the main essay asked about the influence of Italian art on Northern European artists in the later 16th century. How could my superstudent (and supergossipy) friend have been wrong? Worse yet, what was I going to write, since I had consumed most of my study time on (you guessed it) Spranger and his sources? Luckily, upon closer (and calmer) reflection, I realized that the question *was* about Spranger, just under a different description—Spranger just was one of the leading Northern artists influenced by Italian artists. My faith in humanity (and in gossipy friends) had been restored, and I went on to write a fine essay.

OLD ATTITUDE: "Time marches on—with you or without you."

NEW ATTITUDE: "I am the master of my time."

Some students let time get the better of them in an exam. They start off with question one, spend as much time on it as they need, then go on to the next question, take as much time on it as they need, and so on and so on. They never check on the time. They don't even bother to wear a watch. They live in a timeless universe. They have absolutely no idea how much time has gone by, or how much time remains until the professor announces there are five minutes left. Then, of course, the truth sets in. The test is about to end and they have only about half of it completed.

When you are taking a test, don't let time rule you. Take control of it. Track the time and prioritize how you are going to spend your time. Here are some strategies you might use:

- Spend the most time on those questions that carry the most points. (Getting the most points is job 1.)
- Don't obsess and spend unlimited time on some 2-point problem that "I'm going to get right, if it kills me." (It just might, or at least kill your score.)
- When a number of questions are equally weighted (and are independent of one another), try to complete as many as you can before turning to the most difficult of the lot. (Gather as many points as you can, in as short a time as you can, before investing massive amounts of time in that killer problem.)
- When you get stuck on a problem, consider moving on to another. (Your mind keeps working on the old problem while you work on the next, and in any case it is pointless—and depressing—to thrash around on a problem that you might not get right, anyway.)

Most of all, make a conscious plan at the outset of how you're going to apportion your work time. And don't deviate too much from the plan, no matter what the (seeming) emergency is.

OLD ATTITUDE: "I'll set things up nicely."
NEW ATTITUDE: "Set up? No time for a pregame show."

For some students the start of the test is a time to set up. They write out elaborate outlines. They compose very long introductions. But then they change their minds. "What I have isn't quite right," they muse. So they cross out what they wrote and start again. Maybe they start again on the same question or maybe they hop to a different one. By now things are looking nicely set up. Unfortunately a big chunk of the exam time has already passed.

Let go of the set-up mind-set. The start of the exam is the time to get down to real work. Don't be wasting any of your time. This means no outlining—professors never read outlines, so any outlining is a waste of effort and paper. A brief list of points to yourself is okay, especially if it helps you remember and organize your thoughts. But more than a minute or two jotting down acronyms or mnemonic devices is a waste of time.

Make your first paragraph—indeed your first sentence—a clear statement of what you're going to argue or cover in your answer. That way you'll start earning points from the get-go. Perform the following thought experiment. If the grader were to grade *just* the first sentence, would he or she have a pretty good guess of what grade to give me for the rest of the answer? And would it be a good grade? (If not, pick a more powerful first sentence.)

Be resolute. Never stop and start over. If you have a choice of questions, take a few minutes to make a judicious choice and then don't second-guess yourself. Hey, life is short. And the exam is shorter.

How to Choose When There's a Choice

Sometimes your professor gives you a choice of questions to answer. It might be a small choice—say, defining five out of six terms. Not much at stake here. Quickly choose the ones that strike you as best. But sometimes you're given a real and important (*translation:* grade-affecting) choice to make. You're told: pick one of two essay questions, the essay counts 70 percent of the test grade. Here it's worth thinking out which choice is likely to net the best grade.

While generally professors try to make the questions equal in difficulty, sometimes they screw up. So if a question seems much harder, perhaps it *is,* and most of the students who pick it will do badly. Select one of the others. Or sometimes there's a less challenging question, which pretty much everyone can get a B on, but few push hard enough to get an A. If you pick *it,* make sure you go that extra mile to really kill it.

Some choice strategies are clearly bad. You'd have to be a fool to

- automatically pick the first choice offered, figuring it's not worth wasting valuable time reading the rest.
- pick the topic you think the professor *likes* best (rather than the one you *know* best).
- immediately discard one of the questions that asks you to do four tasks, simply because nothing occurs to you to say on *one* of the four parts.
- select a question on the *area* or *topic* you prepared, without regard to the specific *question* being asked from that area or topic.

> • reject a question simply because it's longer than the others, or includes a quote or introductory paragraph as part of the question.
>
> Always make the choice that enables you to do your best work. And to get the best grade.

OLD ATTITUDE: "Why explain? The professor already knows the answer."

NEW ATTITUDE: "Who cares what the **professor** knows? I'm here to show what *I* know."

It generally takes a lot of effort to answer test questions clearly. We mean really clearly, so that someone who was unfamiliar with the issue beforehand could form a clear conception—just from what you say—of what the answer is and what the reasons for that answer are.

Write for a reasonably intelligent person—not a professor (or TA). Think of how you would explain the material to someone somewhat smart—your parent, friend, roommate, kid, whatever—but who has not had the benefit of the course. Take the time to explain (and, in a longer essay, to set up and motivate) the various points you're making. Explain. Really explain. Don't gesture or hand-wave.

OLD ATTITUDE: "Let's keep it short and sweet."

NEW ATTITUDE: "When was the last time you saw a short and sweet **peacock**?"

It's certainly true that everything you say on an exam can and will be used against you in a court of law. Or at least, the court of grading. Make a mistake and it's likely you'll lose some points—and if it's a really big gaffe, lots of points. Given this, it seems that basic conservatism would be the order of the day. You might adopt as your maxim: Write as little as I need to answer the question. Give the bare facts, but nothing more.

This attitude can avert some damage, but it's also a sure way to keep you out of the running for an A. The professor thinks: Bare-bones answer equals bare-bones grade (which in many cases is a B—see Chapter 2). If you want to get an A, you're going to have to *strut your stuff*—to show your professor what you've really got, by providing nice and full answers to the questions asked. Often the professor will give some indication of what length (and depth) is wanted—either by the time allotment for each question or by the space allotment for your answer (if you're being asked to write your answer on the actual test paper itself). Strive to fill the time and space parameters suggested by the professor.

You might be able to give an okay answer in less. But you're not there for the minimum. Don't forget you are strutting. And that by unfurling all your feathers (at least all your relevant feathers) you're going to set yourself above the rest of the muster. (For you non-ornithologists, "muster" is the term for a group of peacocks. Also called a "pride.")

Usually, it's flashier to pick the really best points you can make in answering a question—the ones that will really knock your professor clear across the room—rather than a whole slew of lesser ones (the good, the bad, and the ugly). Then spend your time developing these few, carefully selected high points. Maybe you could give some extra examples or illustrations to support your points. Maybe you could argue more fully and forcefully for your position. Or perhaps you could offer some specific analysis in support of your points (as opposed to covering lots of points more superficially and descriptively). Whatever the case, when you're giving things the quick "once over lightly" you're not strutting your stuff. A point not lost on the professor when he or she evaluates your plumage.

✓⁺ EXTRA POINTER

> The one time you may want to keep your answers short is on a foreign language exam. Even though it would make sense to reward students who attempt more complex answers, thereby using more sophisticated grammatical forms, foreign language teachers (especially in lower-level courses) usually just take off points for every mistake. So here's a place where keeping it short and sweet might actually turn out to be a good idea. Ask your TA or professor how he or she grades.

OLD ATTITUDE: "This isn't all that neat, but so what; the grader will puzzle it out."
NEW ATTITUDE: "Pig mess? A guaranteed C."

Some students turn their test paper into a total disaster zone. They've written things in the wrong place, in the wrong order. Their essays jump around from one zone to another, with arrows to indicate what follows what. They've crossed things out. Written with pens that smear or that don't have enough ink. And if that weren't enough, they write in a way that would take a master cryptographer to decipher. Or maybe a psychic. It's hard to give a paper a good grade when you can't read it. (Sometimes students even hide their answers in parts of the exam book that the professor finds only after he or she has assigned the grade. The professor might even go on to read this hitherto-undetected extra material, but without any real plans to make adjustments to the grade.)

Professors'
Perspective

Have you considered what color ink to write with? The right answer is black, which connotes authority, responsibility, and professionalism. Exactly the qualities a professor would expect from an A student. The worst choice? Felt-tip green or purple. And while you're at it, don't put little circles over your i's or emoticons at the end of your answers. You might think all these things are kinda cute, but professors aren't so interested in your self-expression when it's 11 at night and there are still 39 tests left in the stack.

Have mercy on your grader. Clearly indicate which question you are answering. If a question has subparts, answer them in order. And mark the parts with headers, so that the grader will be able to match your lovely answer to the stock of points he or she has reserved for each part of the question. For problem-solving tests, show all your work. You can often get partial credit for work done. Sometimes the professor will even disregard a small error on the basis that, given the work shown, no one would be stupid enough to make *that* mistake.

OLD ATTITUDE: "I'm going to get out of here ASAP."
NEW ATTITUDE: "Is that Krazy Glue I just sat on?"

Who would want to stay at an exam any longer than necessary? Especially when the exam is tedious? Or the seat is uncomfortable? Or your hand is tired? Or you've got four other exams that day? In high school you had to sit till the time was up. But most college professors don't chain you to the chair. You're free to leave whenever you want. And you'll garner glances of admiration from your cohorts when they see how you've finished the exam in record time. So why not take advantage of your newfound freedom and dazzle your friends as well?

Because you want good grades. And your best shot at top grades comes when you view yourself as glued to the chair until the test is 100 percent over (and then some). That way you can take advantage of all the available time. When you come in with the attitude that you'll be staying until the bitter end, you will be less tempted to give rushed answers. Or to miss questions by accident. Or to make silly mistakes. After all, you've committed yourself to staying the entire time—no matter what. You'll have time to read your work—the way your professor or TA will—before you hand it in. Perhaps a point can be explained better or amplified (sometimes even a sentence penciled into the margin makes a big difference). Sometimes you find you've left out a step in the calculation. Or a verb in the Russian sentence. Or the cultivar, or the ecotype, or the hybrid of a plant in your horticulture test. (Thought we only knew about art history and philosophy? Hey, we've had distribution courses.)

We've often noticed that the exams that get turned in first usually get the lowest grades, while the A exams tend to be the ones handed in last. We don't know which causes which—the quick finish or the bad work. But we do know that working for the whole test period is going to work to your best advantage.

The test is a major grade-bearing moment. Why trade the chance at a really good grade for an extra cup of coffee or a slower stroll to your next class?

OLD ATTITUDE: "No mere mortal could do this."
NEW ATTITUDE: "My middle name is Zeus."

At some point in your college career you are going to face test questions that are tough. Really tough. Even though you've studied. Even though you came to every class, did all the homework, and kept perfect notes. (Hey, you're half-way to heaven.) How could this happen? In some cases, it's simply because you're taking a course in a challenging field. Or have a killer professor.

But lots of times, it's just the system at work. Many professors deliberately construct exams with questions of varying levels of difficulty, so as to better distinguish different levels of performance among the students. When Lynn was giving computer-graded multiple-choice questions in a big lecture class, she would get printouts from the computer center telling her the percentage of students getting the right answers on each question. When Lynn constructed future tests she would make sure to include some questions that had an 80 percent correct answer rate, but others that only 50 percent of the students could successfully answer. That ensured the proper distribution of grades. But it also ensured sweaty moments for half the class.

The upshot? When you get to a tougher question on an exam, bear down. Don't panic. This is the way it's supposed to be. This is the moment when the A students are going to be sorted out from the rest of the pack. Work slowly and carefully. Keep your focus. If the question includes *subtasks* or *subparts*, break the question down into these smaller components. Locate any problem you're having and isolate the difficulty. Often a single insight into a specific point of difficulty can salvage the whole problem.

Jeremy Remembers When . . .

I was taking what I thought would be a relatively manageable mathematical logic class. And I was doing pretty well at it, too: good grades on the homeworks, the tests, could follow the lectures, and so on. I got to the exam, and there were 15 proofs to do. I tried as best I could to complete each problem, but try as I might, there were 6 problems I just couldn't finish. I bore down, focused, twisted my mind. And twisted some more. Still, I left the exam very despondent. I figured: didn't finish 6 out of 15. Forty percent of the proofs didn't close. A 60 percent would be the best I could get. And in a required course, too. Turned out only two students had succeeded in completing those proofs. So while I got a 62, that came out to an A–. Moral? Never say die. At least not during the exam.

Make a directed mental search to relate the difficult task to materials studied in the lectures, discussions, and readings. Try the best you can to forge at least *some* answer (even if not the 100 percent perfect answer) to the part that escapes you. Sometimes when professors deliberately include some harder task in a question, they expect only a partial answer (or beginning of an answer) to that part.

OLD ATTITUDE: "Don't ask, don't tell."
NEW ATTITUDE: "Hey, it's worth a shot."

What should you do when you're *really* stumped by something on a test? Maybe you aren't sure if the question is asking this or asking that, or maybe you don't understand some word the professor is using. Or you don't know if some material should (or should not) be included in your answer. Perhaps something is just totally confusing you.

You might just think that there's nothing you can do. You have no one to turn to—after all, you're not allowed to talk during an exam.

But wait—have you forgotten your faithful professor, who is there hoping that everyone will do well on the test? If something is stumping you, it's always worth taking a shot and asking the professor (or TA) for help. Go up to him or her and quietly raise your question. Do not ask the professor to *give* you the answer. This will cause any professor to immediately get his or her hackles up and send you back to your seat no better off than before. But professors can respond well to comments or questions like "I'm not sure if I should include a discussion of _____ in my answer to this question" or "I was thinking of answering the question in this way—am I on the right track?" or even "I'm very sorry, but I'm not sure I fully understand what this question is asking. What exactly did you mean by _____?" (Be advised that while professors will not usually define technical terms that you are expected to know, they will often explain ordinary English words that are not part of the content being tested.)

So take a shot. Ask away. What'cha got to lose?

OLD ATTITUDE: "Time's beginning to be a factor."
NEW ATTITUDE: "It's not over till it's over."

As the clock runs down on the exam, you might find yourself thinking all is lost. There's barely any time left and you still have questions, or parts of questions, you haven't answered. But it's never too late to take steps to add more points to the board.

Lots of times professors don't end the test on time. They will often allow the class an extra five or ten minutes, especially if there's no class rushing in right after theirs. Don't turn your exam in at the exact second time runs out. See if you can squeeze a few extra minutes out of your professor.

Even if you can't garner extra minutes, you might still be able to salvage an essay by throwing a Hail Mary pass in the final seconds. Go straight for the goal line. Simply *state* the highest points, even if you don't have time to develop them. In some cases, the professor is just looking for a specific point to come up in the essay and will credit you just for getting it in. Put in answers to all parts of the question, even if you have to be very brief. For mathematical and other problem-solving tests, write out what strategy you *would* have used if you'd had more time to finish the problems. And for multiple-choice tests (you guessed it), blacken all d's. (No one is quite sure why this works, but it's usually best to pick all the same letter, and to make your choice the last letter, either d or e.)

Don't give up while there's still some time. Professors are sensitive to the time pressures students face on exams. And everybody likes to see that Hail Mary pass end up in a score.

✔⁺ EXTRA POINTER

Never write an excuse or apology into your exam. Comments such as "Sorry, ran out of time" or "Would have done better, just didn't have time to study" draw the professor's attention to weaknesses in your answer. Which sometimes the professor hadn't noticed or hadn't taken points off for. And some professors could view such remarks as attempts to gain sympathy (at the expense of other students). Let your answer speak for itself. It is what it is.

Review Session

Doing well on tests isn't just a matter of how well you've prepared. It's also a question of what attitudes, preconceptions, and mind-set you come to the test with. Negative and defeatist attitudes—such as that the test will be like the Spanish Inquisition, or only an opportunity to screw up—need to be replaced with positive and constructive views, for example, that the test will be a productive working session for which you are equipped and able to do really well.

At the exam itself:

☑ **Don't be overpowered by first impressions.** You can mess up a question that strikes you as old hat, and you can excel at a question that grabs you by surprise. Take the time to determine what the question is really asking, and what resources you have that bear on the question.

☑ **Forge an initial plan of action and of time management.** Be in control of your time, make an intelligent and reasoned choice of what question to answer (if there is a choice), and get right down to writing your answer.

☑ **Craft your essay nicely.** Be sure to explain fully (not just for a professor), strut what you know (highlighting your very best points), and make it as easy for the grader as you possibly can.

☑ **Don't panic toward the end.** Realize that some questions are meant to be harder, that in many cases the professor (or TA) is willing to help, and that even a small amount of remaining time can be used profitably.

Sure, you need to know your stuff. But you also need to *show* your stuff. Which you can do only if you have the right attitude to test taking.

CHAPTER NINE

The Hidden Value in Going Over Your Test

Ours is an age of interactivity. We challenge you to go to any café or student center and find someone who isn't connected on some device or other. But did you realize that your professor is trying to connect with you every time he or she grades a test? That that same professor who holds court in front of the 200, 300, or 500 assembled students would like to have a meeting of minds with you—one-on-one. Oh sure, your professor isn't communicating with you by e-mail or instant messaging—and isn't visiting your blog, chat room, or user forum. He or she is doing it the old-fashioned way (at least for now), by writing comments on your exam. Still, it's a real attempt to interface with you. In many cases, it's the only attempt the professor will make. Usually there's no connection made. It's too bad. Because making contact with the professor—by going over the exam comments on your own, then seeking out the professor during his or her office hour—is one of the very best ways to boost your grade in the course. If done right . . .

TASK #1: Decide to Go Over Your Test

One day, a week or two after you take your test, you find yourself sitting in class. The professor is giving a fine lecture (okay, an adequate lecture), the class has about 5 or 10 minutes left, and the only thing weighing on your mind is what to have for lunch. And then it happens. All of a sudden the professor brings the lecture to an abrupt end, pauses, and somewhat hesitatingly announces, "I have your exams to give back." You immediately feel your heart beating faster, your stomach clench. It's hard to take in what your professor is saying now—maybe some general words about how the class did overall, maybe how people did particularly well or will need to work harder on this or that, maybe how one class did compared with another. Then the professor hands the exams back, and all of a sudden you're sitting there with the exam paper in your sweaty hands.

You peel to the last page and take a quick glance at the grade. You feel another surge of emotion. Could be relief or pleasure. Or disappointment. Or even shock. Whatever the case, class is over. Time to go. You don't have time to read over the corrections and comments. You just shove your exam into your notebook, throw your notebook into your backpack, and that's probably the last time you'll ever look at the dreaded exam.

And who can blame you? If you did well, there's really no need for further thought. You made it to enough of the classes; you studied enough and in the right way; you got your answer out in spite of the oppressive test conditions. And you got your A (or A– or B+ or whatever you're content with). On the other hand, if you did badly, the last thing you want to do is relive the experience of the test by reading all the things the instructor thinks you got wrong, and, worse yet, looking again at all the things *you* think you messed up. No—out of sight, out of mind. Maybe next time things will turn out better.

There is no even minimally sentient being who doesn't have these feelings about going over their exam. Perhaps amoebas aren't upset when they get back their tests. But for all the rest of us—including all your fellow students who are shoving their exams into their backpacks even faster than you—we'd rather jump off a bridge than read over our tests.

But suppose we told you that by reading over your test you could improve your grade by a third of a grade on the next test? Would you then imitate the amoeba? How 'bout half a grade? Or a whole? Or more than a grade?

We're here to tell you that even though it's one of the least pleasant things to do on the planet, you should make it your business to spend some time going over your exam results. Because this activity—which, by the way, doesn't require more than an hour or so of your time—is one of the more important things you can do to improve your grade. Or, if you did well, to have continued success in the course.

Going over the test can help you improve skills that are needed in the course, and that aren't going to go away anytime soon (this can be incredibly important in any course that is skills-based or is cumulative). For instance, if you go over your test and figure out the organic chemistry problem you missed, you'll put yourself in a position to solve the more complicated problems coming down the pike. If you bone up on the subjunctive, you'll be better positioned to figure out the doubly embedded subjunctives on the next test. If you learn how to do two-variable analysis, you'll have less trouble when they test you on—you guessed it—three-variable analysis.

✓⁺ *EXTRA POINTER*

> If it happens that you aren't in class the day the exams
> are passed back, make sure to pick up your exam at
> the next class or at the professor's office. Professors
> are left at the end of the semester with heaps of pa-
> pers and tests that for one reason or another have not
> been picked up. All these graded tests represent not
> just something to be tossed in the recycling bin, but a
> valuable resource that some students have just thrown
> down the tube.

Going over the test is also of great value in helping you
improve your study techniques. Professors are creatures of
habit and usually repeat the same kinds of questions—just
with different subject matters—on future tests. So going
over the test with an eye to how it was constructed, what
level of difficulty it was on, how the point values were dis-
tributed, what the professor or TA took off for, and so on
will put you in a position to do not just A+ but A++ preparation next time. Which of course will result in an A+++
exam.

But what really happens when you bite the bullet and go
over your droppings (sorry, exam) is that you allow yourself
to experience genuine intellectual growth. You simply learn
better by seeing a single mistake corrected than a hundred
check marks at places where you got it right. Especially if
the mistake occurred at a particularly strategic point—for
example, where your mind was centrally coming down on
some issue, or applying a key concept, or drawing an important inference or conclusion. Seeing a mistake that you
yourself—yes, you—have actually made can be one of the
most important steps in your mastery of a subject or in your
more general intellectual development. You now know

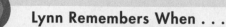

Lynn Remembers When . . .

When I taught a large intro course, I often had students coming to me unhappy with their test results and wanting to do better on the next exam. Because of the large size of the class, I needed to figure out a way to give these students the most help I could, in the shortest amount of time. So I devised a system in which I would spend 10 or 15 minutes with each student, going over two or three questions that he or she had missed on the test, and perhaps making a suggestion or two for next time. When I tracked how these students did on the next test, I was amazed to see that most of the students I had met with got markedly better grades. Some had gone up a full 20 points. Okay, it wasn't a double-blind study. But there really did seem to be a clear payoff for going over the exam.

exactly what piece you were missing. And you begin to see your college experience as a continuous learning process, not just a series of hurdles to be mindlessly jumped over.

TASK #2: Get in the Mood

Like any relatively unpleasant experience in life, going over your test isn't something you can just do. You've got to get yourself into the mood. The right frame of mind. No, you don't need candles, nice music, or a bottle of wine. What you do need is a relaxed and calm state of mind, one in which you can objectively and dispassionately go over your work (and how it was received). Don't plan to go over the test the day you get it back and are still experiencing the thrill of success, the agony of defeat, or something in between. Wait a couple of days till the initial reaction has subsided and you're able to revisit your test with some equanimity. (On the other hand, don't wait so long that you can't at all remember what was on the test, or you're so busy cramming for the next test that you simply have no time for your old one.)

To get into the right mood (and stay in that mood as you are reading through your test), step back from whatever emotions you might have about your test. Then step out of yourself. Pretend that you aren't you. Imagine that you are being asked to evaluate someone *else's* test. And that your job is only to figure out what that *other* person (now no longer you) did right (and less right). And perhaps also to make a suggestion or two—if something occurs to you— about what changes (or strategy adjustments) could be made for next time.

Of course, your primary goal would be simply to *understand* what the grader said: what comments he or she wrote, how he or she calculated the grade, which items counted and which did not, and what good points (and what bad points) the grader took notice of. *In such a situation, you're going to give the grader the benefit of the doubt.* You're going to assume that the grader had *something* in mind when he or she spilled all that ink in the margins, at the end of paragraphs, and at the very end of the test. In every case you're going to try to figure out what that something is. When the grader barks, you're going to think, there must be something he or she is barking *at*.

And you're not going to assume that the grader somehow had it in for the writer of the exam—or somehow made grading errors. Why would you think that? You don't

Jeremy Remembers When . . .

One time a student came into my office, claiming he hadn't understood any of my comments. *"Any* of them?" I asked, incredulous. "Surely you must have been able to understand *some* of them." "No, almost none," the student confidently asserted. I took the paper and started reading the comments out loud, one by one. After about nine comments the student said, "Now I see what my problem was. Those two linked letters at the beginning of many words were 't' and 'h.' All those words I couldn't read were 'the,' 'this,' 'then,' and 'therefore.' No wonder I'm doing so badly in your course."

know either the grader or the recipient of the grade—so why assume there's some animus or bad karma between the two? And since you know that the grader has probably graded hundreds of papers (indeed thousands, if you're at a large state university), you're pretty sure he or she knows how to grade. And is applying some scale, and treating all the students fairly.

Most important of all, you'll really try to read—and understand and think about—what the grader is saying. Even when the writing is messy, or small, or in light pencil. And why not? You're so mellow, so emotionally uninvolved, and so eager to help the exam writer (in reality, yourself) that you're willing to offer the grader any consideration you can.

TASK #3: Go Over Your Test on Your Own

Now that you're in a state of perfect calm—and on the best of terms with the grader—it's time to begin the "going over" process. By yourself. All by your lonesome.

You Can Do It!

Really.

What you need to do is to work through each part of the exam separately—multiple-choice, short-answer, essay, problem, whatever. Without jumping from one part to another and back again. As you go over the exam piece by piece (in much the same way as the grader encountered your exam), take stock, take inventory. Perform a reality check.

Here are five things worth considering:

1. The Lay of the Land

Every test has a certain geography. There are places visited—questions asked. A route from place to place—a certain structure to the exam. Landmarks and points of interest—concepts tested and ideas probed. As you read through your test, consider the lay of the land. How was the test organized? Where did the professor take the questions from? What percentage came from the lectures? From the readings, discussions, or labs? Were the questions ones you anticipated or prepared, or did they seem to spring from thin air? Was the level of difficulty about what you expected, or had you significantly underestimated (or overestimated)? Did you succeed in isolating the main concerns of the course—the professor's main concerns, we mean (those are the ones that count for the grade)—or did you make some misjudgments about what was most important, and what less? This is really the first moment in which you have a concrete touchstone by which you can gauge how well you are *taking* the course. Use it well.

2. What Went Wrong

Any student's test has mistakes. Except if you got 109 percent. As you read through your test, make a directed search for the various places at which you lost points, and figure out what the right answer really was. This time, of course, you can use your books and notes. Maybe the professor gave you an answer key you can consult. Or maybe the very problems—or variants of them—were part of a homework or a problem set that you can look back at. Reflect also on *why* you lost the points you did. Did you miss the class in which that topic was discussed? Did you pass on a reading? Was it that you didn't understand (or didn't remember) some key fact or concept? Or did you really know the answer but blew it because of the stress of the exam?

 IT HAPPENED ONCE . . .

Lynn posts the answers to the multiple-choice portions of her exams in a glass display case in the narrow hall in front of the department office. Students get their own answer sheet back on a printed form, then are "invited" to go over their test, standing up, at that display case—while dozens of students are rushing by to get to their next class. Not too pleasant. Still, if you want to improve your grade in Lynn's intro Art History class, standing there like a giraffe in the midst of the advancing hordes of wildebeest is one of the very best ways to improve your grade. You gotta do what you gotta do.

3. What *Almost* Went Wrong

Don't just look over the questions you got wrong; check over the ones you had trouble with, but lucked out and somehow got right. You know which these are. The ones where you "sorta, but not really, knew the answer." Where what you wrote down was only slightly better than throwing darts. Where you got it right, but for all the wrong reasons. Now is the time to be sure you really understand *why* your answer was right (and the other answers were wrong). Because your choice of answer could have been based on a misconception that didn't hurt you this time, but could come back to bite you next time.

4. What the Comments Really Say

The comments written on your test are the true gold mine of helpful information. The mother lode. The point at which whoever graded your exam—the professor, TA, or some other grader—confronts directly what you say. Often with surprising focus and very good detail. It's the point at which there's a true rubbing up of minds. If only for a brief instant.

On multiple-choice, true-false, and short-answer questions, the comments are usually pretty brief and self-explanatory. A ✓ or an **X**, a score of 4 out of 5, a short code word pointing you to what should have been in the answer. But for essay questions, or more complicated proofs or problems, there can often be quite detailed, individualized comments. These usually fall into two basic kinds. There are **marginal notations**—comments that the professor (or other grader) scribbles in the margin as he or she is reading through the test. These are typically reactions to specific claims that test writer is making—a sort of running stream-of-consciousness commentary that the person reading the paper is writing down, both to correct any mistakes and simply to keep track of the grade as it is emerging through the various paragraphs. Then there are the **summary comments**—things the professor (or TA or grader) writes down at the end of the test. These represent the professor's considered judgment about the strengths, and weaknesses, of the work as a whole. Often these summary comments incorporate (if you know the code) the professor's reasons for assigning the grade he or she chose.

Sometimes you find lots of comments on your test, sometimes only a few. Depending on the style, concern, and workload of the person who graded your test. But regardless of the number, there are some types of comments that almost always correspond to the things that really count for your grade. So as you read through the comments on your exam, be on the lookout for anything that sounds even remotely like any of these:

- Any comments that point out specific parts of the reading, or specific points in the lectures, that you "might" or "could have" referred to or included in your answer. "Might" or "could have" in this context means "should have"—if you wanted to get an A.

- Any comments that make any sort of reference to a mistake. Here look for words such as "No," "Not right," "Inaccurate," "Misconception"—or, if you have a somewhat testier professor (or TA), big X's on the page. Essays often lose points at very specific places. So anytime there are explicit indications of this sort, it's likely these are (among) the reasons you got lower than a B.

- Any comments that use technical terms, or make methodological suggestions. Lynn often writes, "You need to do more *iconographic analysis,*" while Jeremy frequently comments, "You haven't directed your objection to a particular *premise* or *inference* in the argument." Our history friends often write, "You haven't made enough use of *primary sources* in constructing your answer." Anytime you encounter these sorts of comments about the basic tools, methods, and strategies of the field, you'll need to correct your approach before you tackle the next graded piece of work.

- Any comments that suggest ways in which you could "develop your point more clearly," "explain in fuller detail," or "give a few examples." Here the problem isn't so much some mistake you're making; it's that you just aren't clear enough, or don't go far enough, in presenting your answer. For next time, think about ways you could be a better exam writer—ways in which you could produce a document that will be more understandable and more persuasive to the one doling out the grade.

The Lingo of Grading:
Understanding the Summary Comment

Professors tend to write summary comments the way olive canners sell olives. In the canned olive aisle, "large" means "small," "extra large" means "medium," "jumbo" is "pretty big," "colossal" means "big," and "super colossal" means "now we're talking, a good-size olive!" (No kidding, it's a USDA regulation.) So it is with many professors. Not wanting to suggest to any student that his or her work is less than large, professors and other graders often write somewhat inflated summary comments. Ones that take pains to highlight what's good about the answer, and offer upbeat and encouraging words for next time. To give you a sense of how this goes—and how, as a result, one should read summary comments—here are some typical comments we've become quite familiar with (and the grades that go with them). The comments *we ourselves* write:

- "Excellent essay, well-argued and well-reasoned throughout, especially about _____." = A

- "Very good essay, with a strong treatment of _____ and good analysis of _____." = B+

- "A good job, as far as it goes. You might have considered _____ and probed _____ more fully." = B

- "Basically good job. All the information is there, but a better answer would have looked at _____ and _____." = B–

- "A good effort, but fails to consider _____ and has some errors of varying degrees of significance." = C+

- "Some good moments, but makes a serious error in _____." = C

- "Essay shows little understanding of the material relevant to the question." = D

- "A poor job. Essay shows no knowledge of the material relevant to the question." = F

We don't want to say that all professors use the same lingo. They don't. But any professor (or TA or grader) who's been around for more than a semester or two has his or her own set of "grade code words." Be on special lookout for them.

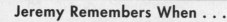

Jeremy Remembers When . . .

You might be wondering how such lingo gets started in the first place. Well, one time when I was a beginning TA (and a good deal less experienced), I met a student in the hall who had recently gotten back a C on her essay test. The student asked why I'd given her the C. I tried to avoid answering, but she kept on asking, so finally I just said, "Well, to tell you the truth, the reason you got a C was because most of your essay was B.S." The student hit the roof—and launched a complaint against me not only to the professor of the course but to the department chair as well. That's when I learned two important lessons: (1) Never talk to a student about grades in the hall, and (2) Always preface your C grades with the words "Some good moments, but . . ."

By now you've worked your way through the morass of comments. You've deciphered the special code in which professors often write. You reduced "puffed up" comments to their proper size. But now we come to what's most important of all. The "cash value" in all the travail. Figuring out . . .

5. Strategies You Can Use to Do Better Next Time

As you go over your test, it's important to be constantly strategizing about things you can do to get to the next grade level (or higher).

Sometimes there are improvements that are no-brainers—ones that would involve very little time investment and net boatloads of new points. If you lost lots of points on simple memorization questions, an hour or two in front of the flash cards might resolve the problem altogether. If you missed because you hadn't captured the relevant point(s) in your class notes, a look back at our Chapter 5 might help you improve your note-taking techniques (without any additional time commitment, either). If it turns out you were confused about a single key concept or application, a 10-minute discussion with the TA or prof might be enough to take care of the problem—and put you in good stead when the very same concept is appealed to on the next test or paper. And if, upon reflection, you were simply too nervous or frazzled to excel at the test, by all means consider again our Chapter 8 and incorporate its test-taking techniques into your game plan for the next exam.

Most often what emerges is that there are only a couple of areas causing the problem—and yielding the grade. If you can pinpoint them by going over your test, you can take directed steps to fix them for next time. Most students we see in the office are not missing by a mile: usually it's just a few adjustments here and there. Doing a reality check by going over your test usually reveals exactly what tweaking is necessary.

TASK #4: Go See the Professor (or TA)

Once in a while, though, things aren't so simple. In spite of your best efforts to get in the mood and go over the test on your own, you just can't figure out where you went wrong. And why you got the grade you did. Maybe the professor didn't really say. Maybe the couple of scribbles in the margin and the one-sentence summary just don't add up to the grade you got. Maybe you just can't figure out what a more complete or more correct answer would *be*—and what exactly was wrong with what you wrote. And maybe (even if you see where you went wrong) you have no idea what you could do differently for next time. You see your grade, and even understand the comments—but no strategic changes leap out at you.

In all these cases you need to go see the professor, TA, or grader. Surprisingly enough, he or she wants to see you.

✔+ EXTRA POINTER

It's always best to get advice right from the horse's mouth. Just make sure you're going to the right horse. If your test was graded by a TA, be sure to talk to *that* TA—not some other TA you might like better, and certainly not the professor. Who might not have graded a single exam in the class—not even yours. If you're unsure who graded your exam, be sure to ask. (And expect an answer. It's your right to know.)

All professors (and most TAs) are required to hold weekly office hours as part of their job. Typically a professor or TA will set aside three or four hours a week, at which time he or she is available to meet with students in his or her office (which may or not be in the same building as the class). The times and place are usually on the syllabus, and are also available on the professor's office door and at the departmental office. Usually no appointment is needed, and unless your instructor is cutting some corners on the job, he or she should be there with an open door. (Sometimes professors even set aside extra "office hours" by appointment. If the professor strikes you as especially nice, you might ask for one of these.)

The few minutes with the professor can have great value—a better grade on future pieces of work (not to mention a real chance to learn something). Most important is that you prepare. That you do your part. This is where going over the test on your own plays an important role—even if you plan to see the professor (or other grader). After going over the whole test at home, select two or three major questions that will be the object of discussion during that office hour. It's *your* meeting—to discuss *your* work. So you should come in with something to say. An issue to talk about or a problem to solve. Professors like that. It shows you've been thinking about the course. And for the professor it breaks the boredom of having to talk about the same question with perhaps dozens of students.

Focus your questions on those items that reflect the central concepts and bear most on the grade. If you tie your professor up with a million nitpicky questions, starting from his or her first comments and moving on through to the last, you'll get very short answers to many unimportant questions. Or a few longer answers to the first couple of points you ask about. In neither case will you have made any meaningful strides toward what you came to find out— where your exam went wrong and how you can do better next time. And, of course, why you got the *grade* you did and how you can get a *better* one next time.

What Never to Say in an Office Hour

There are five questions that all professors hate to hear. So—no matter how tempted you are—*don't ask them.*

WHAT YOU ASK	WHAT THE PROFESSOR THINKS
1. Could we go over the test?	**1.** Sure. If I had an hour to spare. And wanted to be very bored.
2. Can you tell me why I got the grade I did?	**2.** Not any better than the elaborate calculations and reams of comments you've got in your hand.
3. How come my friend (roommate, spouse, person sitting next to me) wrote basically the same exam but got a better grade?	**3.** You both got *basically* the same grade, he an A and you a B.
4. Is there any possibility the grader made a mistake in grading my test?	**4.** Sure. Yours and 75 other people's. (And there's no way I'm going to regrade all those exams.)
5. I *need* an A (or a B, or a C) on this test. Is there anything you can do?	**5.** Nothing legal.

One of the best ways to open your meeting (after a pleasant greeting and perhaps an expression of thanks to the professor for taking the time to see you) is by saying, "There are a couple of things I didn't understand on my exam, and I was hoping you could explain them to me." Then trot out the questions you have prepared, being sure to allow the professor enough time to *answer* them. The point here is to hear his or her insights, not to defend the work you have done. If yours is the siege mentality, you'll learn nothing. Keep in mind also that sometimes the professor doesn't have your exam answer in the forefront of his or her consciousness. How could he (or she) after reading 49 answers to the same question (or if he or she wasn't even the grader)? So offer the professor a chance to read over what you have written, and to review whatever comments he or she (or the other grader) has written down. Be patient. Sit on your hands for a few minutes.

It's also good to remember that sometimes points take a while to develop in any discussion. Sometimes the professor will say something that, if you take the time to consider it, will suggest something for you to say—in "real time." Which, in turn, will incite the professor to contribute something further. All in service of your learning—and getting a better grade next time. *Pool your resources, and let ideas develop.*

Don't be afraid to ask any questions you might have about specific strategies for doing better. For short answers, answers to multiple-choice questions, identification questions, and the like, you might ask the professor where he or she is getting the questions from (if you don't know). For an essay, you might ask whether it would have been better to treat fewer points in greater detail; whether your answer was of sufficient length, or sufficiently probing; whether you should have given more, or better chosen, examples; whether you in fact answered the question head-on, or the exact question that was asked; whether the structure of your essay was transparent and logical; or whether none of these was really the problem. *Let the professor talk more.* In a good meeting the professor will talk for *at least half* the time (after all, you've come seeking his or her opinion).

Of course, if there are substantive issues you don't understand—if there's a central concept you don't get, or a step in a proof or problem you can't do—by all means bring these up. Be honest about what you don't understand. This is not a time for posturing; it's an opportunity to grow intellectually.

Above all, be sure to go away with what you came to get. The answers, clear and developed, to the questions you asked. If, in the process, you've created some goodwill between you and the professor—or if at least he or she now knows who you are—that's a bonus. Which you'll be able to use later, when you will work with your professor to produce the perfect paper.

When To Dispute the Grade (and When Not)

You might have noticed that in all our discussion about going to see the professor we haven't mentioned what you might have thought was the main reason for making the trip—to dispute your grade. This omission is not accidental. We believe that in most instances grade disputes are simply pointless—and that, in any case, the time spent with the professor could be better used to jointly forge grade-improvement strategies. But if you feel compelled—and sometimes you should—here is a table of grounds for dispute that will always work, that will never work, and that could work (with any luck):

ALWAYS WORKS	NEVER WORKS
1. Grader made a calculational error.	1. Test was unfair (or too long, or too hard).
2. Grader didn't notice a page of my answer.	2. Grading scale was too hard (or unreasonably favored some sort of question, or was inappropriate for this level of course).
3. Grader misread some numbers or variables (in a problem-solving exam).	3. My friend (who studied the same amount, wrote the same exam, knew no more than I, etc.) did better.
4. Grader misidentified the question I was answering. (Can happen when there's a choice of questions on the test.)	4. Grader was incompetent (didn't know the stuff, has never graded before, had it in for me, etc.).
	5. I must get a C to keep my scholarship (to remain in good standing, to stay on the track team, to keep my visa, etc.).

SOMETIMES WORKS

1. Grader didn't fully understand what I was saying. (Best if you can point to what paragraph or sentence he or she didn't get.)

2. Grader simply missed a point I was making. (Best to pick a central point, and to show how no comments were written there.)

3. Summary comment indicates the grader didn't read carefully. (Best to be able to point to the sentence[s] that shows the grader's lack of attention.)

4. I didn't get enough partial credit for the work shown. (Good to argue how the work shown constitutes the core of the solution.)

Visiting Professor

Rich Pattis, Carnegie Mellon University

This incident happened when I was teaching at the University of Washington, in Seattle. During a typical quarter, my students complete 10 take-home quizzes (one per week), which I hand out on Friday and collect on Monday. I drop the lowest two scores for each student, primarily for my benefit (although student averages do rise): to attenuate arguments from students who score a zero because they lose a quiz, or forget to pick it up, complete it, or bring it in for grading.

One semester, on the morning of the third Friday of the term, there was an extraordinary windstorm; all classes were canceled, so I could not distribute that week's quiz. On Monday, I announced that rather than assign two quizzes in the same week, I would drop this quiz for everyone. I promptly received a message from one of my students. He had not bothered to take the first two quizzes, knowing that two would be dropped. Now, by my policy of dropping everyone's third quiz, he would have an unwarranted zero among the eight quizzes I counted.

I found the student's logic convincing, and arranged to record one score for him based on the statistical profile of all his other quiz scores. But I felt compelled to warn him strongly that by "blowing off" the first two quizzes, he was not making good use of my policy.

Review Session

One of the best methods for improving your grade is also one of the least well-known (and certainly the least done)—going over your test. Think of it as a four-step procedure.

1. **Get your mind around the importance** of going over the test: It'll help you improve your skills, perfect your studying techniques, and provide you an opportunity for genuine intellectual growth.

2. **Get in the right mood** for going over the test: calm, collected, and dispassionate, trying to understand what the grader is saying and willing to give him or her the benefit of the doubt.

3. **Go over the exam on your own,** paying special attention to figuring out the lay of the land; seeing what went wrong (and almost wrong); understanding what the marginal notations and summary comments really mean; and devising specific strategies for doing better.

4. **See the professor (or other grader) in an office hour,** making sure that you've prepared two or three main questions, that you structure a frank and open discussion of the issues with the professor, and that you leave with answers to your questions—and a good feeling in the professor.

The Greek poet Sophocles once said, "Wisdom comes only through suffering." He must have known about going over your test.

PART 4

THE PAPER

CHAPTER TEN

Understanding the Assignment

The college paper is a kind unto itself. Unlike other kinds of writing you might be more familiar with—the report or summary, the journal entry or newspaper article—the college paper comes with its own set of rules and expectations. That aren't always told you—by the professor or even by the TA. Luckily for you, we college professors have to churn out academic papers all the time. It's how we spend our time (a whopping 60 percent) when we're not teaching. And we're happy to disclose to you our four-step plan for writing an excellent paper: (1) understand the assignment, (2) do the analysis or research, (3) go to see the professor (or TA), and (4) write that perfect paper. Didn't quite get all that? Read this—and the next three—chapters, in which (you'll be happy to hear) we trace these steps, one in each chapter.

When it's a *test* that's being announced, students often feel a sense of great urgency. When will the test be? What will be on the test? How should I prepare for the test? So too with papers. Only this time, the urgency is to delay. Why not put off addressing this unpleasant task till sometime far off in the future? Like a couple of nights before the due date.

Come to think of it, the night before wouldn't be all that bad either.

Surprisingly enough, this usually turns out to be not such an excellent strategy. That's because top-notch college papers aren't the sort of thing you can just throw together at the last minute. Successful papers are almost always the result of a *process* that takes place over an extended period of time. A process in which you think through—and rethink through—some issue. A process in which your mind continues to work on the problem or question even when you're not consciously thinking about it. This process needs time to develop and gel. There's a big difference between papers that have been developed over a week or two—with the ideas better thought out, more convincingly argued, and better written; and papers written the night before, which almost always have a coarse, rough-and-ready look. Professors are fully trained to recognize this difference in "look" between these two types of papers. And to capture that difference in the grade they mete out to each.

TASK # 1: Get to Know Your Assignment

The first thing to do is to read through your assignment as soon as you get it.

No, we're not saying that you need to start reading the topic(s) as you are walking out of the lecture room. (There are limits even in the *Professors' Guide.* And we wouldn't want you to knock over any of your compatriots, which wouldn't help your grade anyway.) What you need to do—either that night (if you're a night owl) or the next day (if you're a morning person)—is to plunk yourself down and read the assignment. Not in a rush, not skimming, but with some care and precision. As if you were really going to do the assignment.

Professors' Perspective

Most professors actually give considerable thought about *when* to assign a paper. They take into account not only where the assignment "sits" in the course, but also how much time students should spend working on that assignment. They try not to hand out the paper too early, at a time when no student in his or her right mind would bother to begin, or too late, when there really isn't enough time left to do a decent job. *Take the time of the assignment of the paper as a silent clue from the professor as to when to start working.* To begin the process that will culminate in the perfect paper.

Which you are. Survey the landscape. Begin to see what you are being asked about (attending to any choice there might be), and begin to forge some plan (however preliminary) about how you are going to do that task. *Note down whatever first thoughts or impressions you have.* While your first word may not be the last, often these nascent and spontaneous ideas can form the core of what you're later going to argue.

If you're lucky (and it's amazing how often students *are* lucky in this way), acquainting yourself with the paper topic(s) can have an immediate payoff. Lynn, in her Medieval Art course, assigns a paper that asks students to work on the ninth-century Book of Kells. By a fortuitous coincidence (actually, by Lynn's careful plan), she hands out the paper assignment just two days before the class covering—who would've guessed—the Book of Kells! Now, if you had delayed reading the paper assignment (or worse yet, missed the next class), you might not have located the points in the lecture that bore directly on the question assigned. Those little "asides" that shed additional light on "points you might consider in constructing your paper" would have flown by you with barely a notice. All because you hadn't taken the time to preacquaint yourself with the work you were going to have to do in the paper. (Which, by the way, the more "with-it" students had done as a matter of course.)

Look over the paper assignment (and the various choices of topics, if there are more than one) as soon as you get it. So that you not only get the wheels and springs of your mind going (always a good thing), but also gather relevant information and strategies from the lectures—and also readings, sections, and office meetings—that can follow immediately after you receive the topic(s).

TASK # 2: Figure Out the Question(s) Being Asked

Students often come to college thinking they'll be able to choose their own paper topics. Or that they'll be assigned very general assignments such as "Write a paper on Pakistan." It doesn't usually go that way. Most often, students are given very specific questions to work on—often with subquestions to be discussed and/or subtasks to be done. (And even when professors allow—or require—students to pick their own topics, the professor has a pretty good idea of what would be a good question to ask—and what would be a good answer to give—in that sort of course.)

The result, of course, is that it is most important that you figure out exactly what the professor is asking. Not "more or less (but probably less)" or "I sorta understand (but not really)" or "I have a pretty good idea about the question (but I'm a little fuzzy on the details)." No, what you need is a really precise understanding of what each question (and each part of the question) is asking.

Actively determine the *specific* question (or questions) being asked. Distinguish it from *related* but different questions, and also from questions on a different *level of generality*—either broader or more narrowly focused questions. And actively determine the exact *tasks* you're being asked to do in answering those questions. Pay special attention to any verbs in the question: **compare and contrast** is different from (merely) **contrast; evaluate** is different from **explain; state** is different from **argue; sum up** is different from **analyze; raise objections** is different from **develop illustrations**; and so on. Professors spend hours (well, at least minutes) in precisely forming these questions. Spend at least a few minutes understanding exactly what your professor is trying to ask.

IT HAPPENED ONCE . . .

In a recent paper assignment, Lynn asked students to "*characterize* the style of the 17th-century sculptor Bernini's *David*." Not being fully sure that to "characterize" meant to "describe the qualities or peculiarities of" (*American Heritage Dictionary,* p. 226), some students simply *named* or *labeled* the style. This basically sent those papers on a downward cascade (into C land) that could easily have been avoided had the students simply had a firm grasp of what "characterize" meant.

Another thing to consider is whether there is one task or multiple tasks. And if there are multiple tasks, are they truly distinct (and of equal value), or is one the *main* or *primary* task, and the others only *secondary* or *subordinated* to the first?

Depending on how you see the question, you'll think about it—and write about it—in different ways. Also important to consider is the relative *weighting* of the tasks. Does the professor intend that you spend equal mental space—and equal writing space—on each of the topics? Or is one task supposed to take up less thinking—and less writing—than the rest? All of these questions are important to think through from the get-go, not only because your preparation will be influenced by how you take the question, but also because in many cases the percentages of the grade will be allotted based on how the professor intended the question.

In short, *mine* the question. *Milk it* for all it's worth. Search out any *clue*—even a mere shade of meaning or choice of words—that might shed some light on what exactly the professor expects you to do.

But now the most important tip of all: *If at any time when you're puzzling out the question(s) you're not 100 percent—or at least 99 and 44/100 percent—sure of what you're supposed to do*, go see the professor *(or TA)*. *Do not keep working. It won't end well.*

Jeremy Remembers When . . .

I was taking a second-year Ancient Greek class, which focused almost exclusively on translating sections of Homer's *The Iliad*. Midway through the semester, the professor sprung a paper on us: "Write a 10-page paper on anything you want." Thinking that most of the class time had been spent on grammar, I decided to write a paper on uses of the particle "since" in *The Iliad*. Turned out the professor wanted a *literary* analysis of some theme, character, or plot device in *The Iliad*. I got a B– on the paper—which really ticked me off, since I could have gotten an A had I only known what to do. Or thought to ask.

IN OUR HUMBLE OPINION . . .

Ever wish the professor gave you free rein to pick your own topic? To work on something that you were really interested in or really mattered to you. Well, stop wishing lest you get what you wished for. Having to choose what to write on is a particularly dangerous situation—one fraught with risk and the possibility of getting a bad grade. Pick a bad topic, pick a topic that isn't appropriate to the field, pick a topic that's impossible to do, and you're doomed right from the start.

We think it's unbelievably hard for students who have little or no familiarity with a field to pick a suitable topic to write on. (Graduate students, and indeed professors, often pick bad topics in their own research and spend large amounts of time chasing down what turn out to be dead ends.) So if you're given a choice between doing an assigned topic (that is, a designed-to-work topic) and one you have to make up (in other words, a roll-the-dice topic), by all means choose the assigned topic. And even if the professor asks you to pick the topic, you should still have the professor pick the topic. Or at least sign off on the topic you've picked.

Like how? The simplest strategy for enlisting the help of the professor is to go see him or her during an office hour. And just *ask* for suggestions. Many professors (and TAs) will be happy to help. They feel you really care (which you do, 'cause you've shown up). And that they can really teach you something (which they can, at least what topic to work on).

Other professors might be less inclined to help. They might have some purported educational reason for refusing to give topics ("Part of doing the paper is picking the topic"), or they might feel that it's simply not their job to help you find paper topics ("The assignment was for *you* to construct a question. Now go do it"). In this case you should resort to stealth—or at least cleverness—to get a topic. Here are some strategies you might employ:

- Think up three or four topics yourself, then go to see the professor and ask which of these topics he or she "thinks would be the most promising."
- Look over your lecture notes to see if the professor ever suggested things "that might be interesting to do in a paper."
- Look over any study questions or discussion topics that were given by the professor; then ask him or her whether one or other of these would be "suitable for a paper."
- Ask a TA in the course (or advanced undergraduate or someone who has taken the course) what he or she thinks "would be good to work on" for a paper in this course.
- Cry, beg, or prostrate yourself before the professor. Hey, you never know. . . .

TASK #3: Decide What *Kind* of Paper You're Writing

Once you've figured out *what* the paper assignment is asking, the next thing you need to figure out is what *kind* of paper the professor is assigning. Didn't know there were different kinds of paper (beyond first paper, midterm paper, and term paper)? Well, it turns out that there are two main kinds of college paper—the **analytical paper,** in which you are asked to methodically examine some object, issue, or phenomenon; and the **research paper,** in which you are asked to perform some scholarly or scientific investigation or inquiry. There's not a hard-and-fast distinction here. Many research papers could ask you to analyze and evaluate the sources you've located or the data you've collected. And there could be an analytical paper in which you carefully analyze or dissect some given text, or argument, or document, and then confirm your analysis by seeking out (that is, researching) other texts, arguments, or materials— and then analyzing *them*. But the most likely paper assignment is either an analytical or a research paper (rather than a hybrid of the two). Let's take a closer look at the difference.

An *analytical paper* will always ask you to perform an *analysis* of some object, work, phenomenon, or issue. You'll be asked, typically, to consider some complex object of study, reduce it to its component parts, then evaluate the significance of these parts relative to one another and to the whole. A three-part task: study the object, break it down into smaller pieces, and interpret (or assess or evaluate) those pieces.

For example, in a *literary analysis,* you might be asked to dissect the plot, or the characters, or the symbolism, or the theme of some great epic or novel. In a *historical analysis,* you could be asked to determine (and evaluate) the causes of some historical event; to assess how political, economic, and social factors influenced some historical trend or phenomenon; or to interpret some historical document or record with regard to a specific issue. In a *musical analysis,* you might be asked to hypothesize about why some composer added bars to an earlier draft of a piece (and what effect this change had on the piece as a whole). In a *philosophical analysis,* you could be asked to take some argument, break it down into its component premises, then evaluate the success of the argument by assessing its various premises (and their relation to the whole). And in a *financial analysis* (say, for a business or economics course), you might break down some problems facing some multinational corporation, assess the overall situation in light of those problems, and even make specific recommendations about how that company might best proceed.

Though the expectations differ somewhat from field to field, analytical papers typically have six hallmarks. They:

- **Require Analysis:** Critical thought, interpretation, or evaluation are always needed—pure description is never sufficient.
- **Call for Reduction:** Most often, you're asked to dissect some complex object of study into its simpler elements or components. That's because those who do analysis believe that one way to better understand some complex whole is to understand the components of which it is made up (and how they are arranged).
- **Involve Selectivity:** Since you can't consider all dimensions or components of an issue, in virtually all cases you're asked to make a judicious selection of what points are to be studied. (Ultimately, your success in an analytical paper depends to a large degree on just what points you have selected to probe.)
- **Need no research:** You are usually given the object to be studied and are not expected, nor generally would it be useful, to do library or Internet research.
- **Demand hard thinking:** Instead of being asked to think broadly about a wide variety of sources, you're being asked to study—often very deeply and in a very focused manner—a single object, phenomenon, problem, or issue.

- **Use field-specific methods:** You're always asked to apply the concepts, techniques, and methods of the particular field in question to the case at hand. The perspective is always that of someone working in that particular discipline, using the tools of analysis of that particular field.

The other—to many students more familiar—kind of paper is the *research paper*. Here you are asked to investigate, or inquire about some subject matter that goes significantly beyond the materials presented in the class (whether lectures, sections, lab experiments, or readings). You might be asked to read books, articles, or experimental reports written by different scholars. Or you might be asked to dig up historical or literary texts—or perform experiments or interviews or surveys, which you (and other scholars) can then study directly. In either case, research essentially involves collecting and ultimately evaluating new materials—sources or data that have not been presented in class. In every case you need to go beyond your own mind—and that of the lecturer.

Even though the two main types of papers, analytical and research, have some real differences, sometimes it's not as easy as you might think to figure out which you're being asked to produce.

Sometimes a topic could be assigned as either an analytical *or* a research paper. Same question, different expectations. One professor wants you to work on your own, to apply the course methods, tools, and techniques to an interesting new case. Another professor wants you to research new materials, to see how the findings of others can help you understand the texts in question. The fact that it's not immediately clear from the question what the professor has in mind makes it doubly important that you uncover *your* professor's expectations. That is, if you want to get an A (or often, even a B).

POP
QUIZ

Pop Quiz

Consider the following actual paper assignment from an honors humanities course:

What are the main points of similarity and difference between the treatment of the Joseph narrative in the Old Testament (Genesis, chaps. 30–50) and in the Qur'ān [Koran] *(Sura 12)? How does the literary and historical context of each text help explain the differences?*

Question: Is this an assignment for (a) an analytical paper, or (b) a research paper?

Answer key:

Turns out, the answer is (c) could be either. At least for all you know so far.

To see how all this might work in practice, let's take a closer look at the assignment from our pop quiz.

In the Joseph example, the professor could be assigning the question as an *analytical* paper. In this case the prof would be expecting you (1) to examine the two narratives extremely carefully, taking note of not just the similarities but also the differences between the two presentations; *and* then (2) to offer some hypothesis about how the facts about the literary composition of the works, and the historical circumstances in which they were produced, account for these differences.

Here's a place where Task #2 comes into play. In order to do well on this question, you'd have to notice, first, that there are two parts to the question, each asking a different task (find the main points of *similarity and difference*, then *account for the points of difference*). Then you'd have to tease out the structure of the question: the second task is *subordinated* to the first, and the second task asks you to marshal the literary and historical circumstances only in explanation of the points of *difference* (it would be a waste of time and a possible point-loser to devote any attention in the second part of your answer to finding reasons for points of similarity).

Once equipped with a clear view of the question, you'd go on to do the analysis (remember, we're supposing the question is assigned as an *analytical* paper). You'd think long and hard about the presentations of the story in the two scriptures. You might dissect or break down the narratives into meaningful component parts. You might make judgments about which points of similarity—and, more important, which points of difference—are most significant for the analysis you're performing. You might look to the lecture (or reading or discussion-section) material for information on the backgrounds of the Old Testament and the Qur'ān. Any (nonmalicious) professor who expected this kind of paper would have presented analyses of similar Old Testament and Qur'ānic texts (perhaps in the lectures) and would have given enough information (maybe in class or in the readings) on the literary and historical backgrounds for you to *do* the paper without additional input or data. No pilgrimages to Mecca or Jerusalem—or the library or Internet—needed.

But the same exact question could be being assigned as a *research* paper. In this case, you'd be expected to investigate (and evaluate) additional materials in order to develop your answer to at least the second (and maybe even the first) of the two questions asked. You might look at books on eighth century B.C. ancient Israel, and seventh century A.D. Arabia. You might use scholarly encyclopedias or dictionaries on the composition of the texts of the Old Testament and the Qur'ān.

And if you hit it really lucky, you might find scholarly articles (either in print journals or electronic databases) about the very subject of your paper—the presentations of the Joseph narrative in Genesis and in Sura 12. (In the second part of the next chapter, we'll have suggestions galore about how to do just this sort of research.) But whatever the case, the *re search* version of the paper will have to incorporate significant knowledge or information that you do not think up on your own—or even glean from the course lectures, sections, and readings. You will have to be hunter and gatherer—and also processor and evaluator—in your quest for the A.

So how can you *tell* the difference between an analytical and a research paper, if one and the same assignment could be either? Look at the full paper instructions for subtle (and not so subtle) hints. If your professor writes, "This is not intended to be a research paper," or conversely, "Be sure to consider four to six scholarly sources in constructing your answer," this might be a sign of what kind of paper he or she is expecting. Also pay special attention to the professor's oral explanation or any instructions he or she gives when handing out the paper assignment (be sure to write down any additional tips or explanations that the professor offers up— never rely on your memory for these directly grade-affecting tips). And sometimes the whole set of circumstances or the context in which the assignment occurs—what sort of thing the professor has been doing in the course to that point, what sort of activity is going on in section, how the current paper relates to the previous one(s)—can give an indication of what sort of paper is wanted.

But here's the most important (though by now not the *newest*) tip of all: *If at any time when you're thinking about the question(s) you're not 100 percent—or at least 99 and 44/100 percent—sure about what* kind *of paper you're supposed to be writing,* go see the professor *(or TA). Do not keep working. It won't end well.*

✓⁺ *EXTRA POINTER*

In some introductory courses—particularly freshman
writing or basic composition courses—you might encounter a kind of paper called the **synthesis paper.**
This kind of paper, common at community colleges,
junior colleges, and some liberal arts colleges, isn't so
much a distinct kind of paper (as opposed to analytical and research papers), as it is a beginner's paper. A
sort of paper, not written in any particular discipline,
but rather a generic paper, meant to teach you the es-
sentials of writing across different disciplines. If you
encounter this sort of paper, you're in luck. At least
from the point of view of the grade. For the whole
course—lectures, readings, textbook, homework,
quizzes—is geared to teaching you just how to write
these sorts of papers (there's no "curriculum" above
and beyond the writing). So make full use of the
course resources. Attend all the lectures, do all the
readings, go to the writing center, or make use of the
online course resources. And, most important, as always, go to see the prof if you're not sure quite what
you're supposed to be doing.

Review Session

You're off to an excellent start. You've figured out exactly what you're going to be doing on that all-important paper. But just to be sure you're *ready* to move on to the real heavy lifting—actually doing the analysis or research, going to see the professor (or TA), and writing the paper—check to see if you've successfully completed the following tasks. Have you . . .

☑ **Acquainted yourself with the paper assignment**, familiarizing yourself with the choice of questions just as soon as you got them?

☑ **Figured out exactly what the question is asking**, paying special attention to the language of each question and the relations among the subquestions (if any)?

☑ **Diagnosed what kind of paper is being assigned** (analytical, research, or hybrid) and perhaps even had a few thoughts about how you might start that sort of task?

☑ **Thought of seeing the professor**—and maybe even figured out where his or her office is located, and when he or she will be there (if you're especially motivated)?

If so, you're ready to start work in earnest on actually preparing your paper. Turn the page. Posthaste.

CHAPTER ELEVEN

Doing the Analysis, Doing the Research

And now the fun begins. It's time to work, not on the college paper but on your college paper. Not just to think about what the question is asking and what kind of question it is, but to begin to figure out the answer to the question—and to do the analysis, do the research (or both). But you've barely got your mind around what analysis even is. And, as for research, it would be easier to find your counterpart at a Nepalese university (after all, everyone is separated by only six degrees of separation) than to find some monograph or article on that super-obscure topic your professor has assigned. Luckily (as always) we've got hard-hitting tips for you—first our best strategies for the analytical paper, then knock-'em-dead tips for the research paper. Continue straightaway if analysis is your game; skip ahead to page 224 if you'd like to try your hand at research. (Read both if you'd like to see how the other half lives, or you've been blessed with a hybrid paper.)

PART ONE

Top 10 Tips for Doing the Analysis

Here are our best ideas for doing *analysis*:

TIP #1: Work in Your Head

This is one of the easiest steps to take. It requires no action on your part (unless resistance is an action). Resist the temptation to do research. When you feel yourself thinking, "I oughta go to the library, or to the Internet, to find the answer," stop and think, "Don't go there, the answer isn't there." As amazing as this might seem, you're being asked to answer the question or solve the problem *wholly on your own*. No brownie points or extra credit for unearthing and quoting hordes of distinguished authors. No, this is the lonely task of working in your head. Be sure you stay there.

TIP #2: Assemble the Relevant Materials

Just because you're working in your head doesn't mean you have to work with an empty head. You're allowed—indeed often required (at least to get an A)—to import materials from the course in service of your analysis. Perhaps the professor gave some lecture on the object of the analysis, which you can take as the base for your work. Perhaps there was a discussion, or something in the assigned reading that could be useful in dissecting the topic or issue of the paper. You're not supposed to be starting from scratch here. The wheel isn't yours to reinvent in an analytical paper.

TIP #3: Select the Highest Points for Your Analysis

Often when you analyze some text, or article, or object, a number of points will occur to you. You start thinking about some object of study and all of a sudden dozens of ideas spring to mind. All of them good, it seems. And all of them worth pursuing, in some way or other. It's generally good to select the one or two *best* ideas—those that cut most deeply, those that represent the most direct answer to the question being raised, and those from which flow the most important of the other points. Try to at least start in your thinking with those few points that will be most convincing to the reader. Your best stuff. And don't be afraid to hold off in your thinking about those other (somehow worthwhile, but less central) points until later in your thinking. Especially if the paper you're planning out is short, say, four to six pages.

TIP #4: Delve Deeply

Once you've selected the couple of points you're going to focus your thinking on, probe deeply. Try to understand—with 100 percent precision and the focus of a heat-seeking missile—what's being said in a text, what's being depicted in a picture, what's being shown in an experiment, or whatever it is you're being assigned to analyze. Then think about it some more, and some more. Torture the data. After all, since you've limited your focus to only a few main points, you'll have plenty of time and mental energy to scrutinize the object of your choice.

Realize also that your first thought may not be your last. Since you (and your professor) have set aside a reasonable amount of time for you to probe the issue, your ideas can gel—and change—as you think through additional dimensions of the issue. Don't abort the process too soon. Often the difference between a B and an A paper is not so much the truth of the claims (the B paper can often be 100 percent right), but the depth and richness of the analysis. Which is achieved only in time, as you think and revise, then think and modify, then think and tweak (notice that the changes get smaller as you get further into the analysis).

TIP # 5: Look for Relations among Your Ideas

If your analysis is going to culminate in a well-reasoned and logically organized paper, you're going to have to consider the *relations* among the various ideas you're having. It's not too early to think about this as you're starting to do the analysis. Decide which ideas constitute your main points, and which support, illustrate, or expand them. Which ideas constitute your primary line of argument, and which represent your defense of this line against possible objections. Which ideas are direct analysis of the object of study, and which place the results of the analysis in a broader context.

And be willing to play with your ideas. Sometimes an idea that started off as a subordinate idea, marshaled only in support of some more central point, can rise to the level of a primary idea, for which other support will be sought. And sometimes as the structure of the parts of your analysis becomes clearer, some idea can be downgraded, or even disappear altogether. Whatever the case, keep your mind limber and let your ideas evolve naturally, at least in the early and middle stages of your analysis.

WANN'AN (Ⓐ) **?**

> The structure of your paper will to a large extent mirror the structure of your thinking. Look for strategic relations among your ideas as soon as you have them. And keep probing as you continue to think out your paper. Your level of insight into, and your creativity about, the relations among your ideas is one of the keys to getting an A on many analytical papers.

TIP #6: Assess the Relative *Importance* of Your Ideas

It's not especially satisfying to see an analysis in which all the ideas—though relevant to the topic—are assigned equal weight. "Here's a key point," the reader thinks. "And there's another. Then there's this point of capital importance. And that salient detail. And this essential element. . . . Wow, everything's of great importance. Which means that nothing is." In carrying out your analysis, try to develop an understanding of the different *levels* of importance of your ideas. Some are of primary importance, the key claims in your thinking, while others are of lesser importance—still important, yes, just of somewhat lesser importance to your analysis.

If your analysis requires you to provide reasons for a particular event or phenomenon, don't stop at just generating a "laundry list" of equally valid reasons. Don't think of the analytical process as simply locating a series of points or examples in support of some main idea. Think of it as making *judgments* and *assessments* about which of your ideas are most significant, and why.

TIP #7: Follow Your Professor's Lead

Every discipline has its own methods of analysis. As a practitioner of the field, your professor has developed his or her own particular approach to the material (which could be completely standard, totally idiosyncratic, or somewhere in between). The most successful papers in any course are those that adopt the *methodology* of the professor teaching the course—not alternative methods that might be used in other fields or by some other professor. Watch carefully as your professor does analysis in class. Does he or she use a top-down method (starting with a theory and then applying it to particular cases) or a bottom-up approach (looking first at particular cases and then developing a theory)? Does your professor tend to work more narrowly (studying small sections of text, or individual objects or cases, in great detail), or does he or she paint in broader strokes (tracing more general themes through many texts, objects, or cases)? Does your professor interpret cautiously (insisting on collecting complete data before beginning any sort of interpretation), or does he or she interpret more boldly (offering more speculative interpretation based on limited data)? The more you can model your analysis on the kind of analysis your professor likes best, the better your grade is likely to be.

TIP #8: Accept Uncertainty

In most analytical papers, there's no 100 percent right or wrong answer. So there's no 100 percent certainty. There are judgments. And it's up to you to make them. What to put in, what to leave out. Which points are to have more prominence, which are to occupy a lesser role. This is the nature of the beast. Do not let it upset you. (Of course, if you make mistakes, or make bad judgments, that *should* upset you.)

TIP #9: Record Your Thoughts

As you go through the "thinking out" process, be sure to record the thoughts you're having. And not just with a code word or two, scribbled on some tissue. No, a real record that you'll be able to puzzle out—with all the *depth* of your thought—when you later come back to it. (Some A-students even carry around a digital recorder when they're working on an analytical paper, to capture those momentary insights that later morph into key hinges in the analysis.) There's nothing worse than having what turns out to have been a really good thought only to find that it has disappeared forever, either because you didn't bother to write it down or because you simply can't read what you did write. If analysis is a process, it's good to keep a *history* of that process.

TIP #10: Generate an Outline

Your preparation process should culminate in an outline. At the *end*. Only then do you know the relations of the various thoughts you've had. How they all fit together. Which are the central points, which are the derivative points, and which points can be thrown out altogether. A good outline isn't a mere chronology of the thoughts you've had—or even a linear ordering of the points that are going to go into your paper. A good outline is an exposition of the *structure* of your paper. Which you can know only at the end of your "thinking out" process.

But most important of all (as always):

BONUS TIP : Go See You-Know-Who

If at the end of your analysis process you're still not 100 percent—or at least 99 and 44/100 percent—sure about what points you're going to make (and how you're going to make them), *go see the professor* (or TA). Do not keep working. It won't end well. (Heck, you can even go in the *middle* of your analysis procedure if you have a good question or one that could help you along in the process—professors love to be involved in the actual thinking out of the answer.)

Review Session

The analytical paper can be a real mind meltdown or a genuine brain-expander, depending on whether you know what you're doing. Here are our top 10 tips for doing the analysis:

1. **Work in your head.** That's where the answers are.
2. **Assemble the relevant materials.** Analytical papers don't live in a vacuum.
3. **Select the highest points for your analysis.** Your A-1 best ideas, the ones that will knock the reader across the floor.
4. **Delve deeply.** Get to the core of the issue, then dig deeper.
5. **Look for relations among your ideas.** The more interesting, creative, and nuanced the relations are, the better an analytical paper you'll have.
6. **Assess the relative importance of your ideas.** Some more, some less.
7. **Follow your professor's lead.** The methodology that he or she is teaching in the course. And the one he or she will be looking for in your paper.
8. **Accept uncertainty as you do your analysis.** Not being 100 percent certain you're right is an unavoid-

able part of the process. And a sign that it's going well.

9. **Record your thoughts.** As you have them. (You never know what will be important later.)

10. **Polish it off with an outline.** A considered plan. Looking forward and looking back.

Bonus Tip. Go see the professor (or TA). Deal him or her in at the formative stage.

PART TWO

Top 11 Tips for Doing the Research

Here are our best ideas for doing *research*:

TIP # 1: Get Out of Your Head

In any research paper, you are being asked to go beyond your own thoughts to investigate, or inquire about, some problem, issue, or question. If, say, in a zoology class you are asked to study the mating habits of yaks, you could hop a plane to northern Tibet and observe, firsthand, the reproductive behavior of these strange mammals. Such study is what's called **primary research,** since it involves direct observation and study of the phenomenon in question. (Other instances of primary research would be archival investigation of settlement records in 19th-century Iowa [for a history course]; statistical analysis of voter biases [for a political science paper]; and designing and performing an experiment to detect a certain polarization property [for an optics course].) But another, more common research assignment asks you only to study the scholarly literature about those hairy beasts. This is called **secondary research**, because here you study only the (primary) research that scholars before you have performed on their pilgrimages to Tibet. Books, articles, essays, research reports, yes. But no yaks.

Whether your assigned research is primary or secondary, it cannot be performed from the comforts of your room. Regardless of how powerful a laptop you have, and no matter how proficient you are with search engines, you're going to need to get out of your room. To Tibet, or the lab, or the library, or wherever the object of your study lives. In many courses, the first step on any research adventure is the library. You know, that big building with all the books. That's where the answer—or at least the materials for the answer—will be found.

TIP #2: Start Your Research ASAP

Start the actual research right away. In addition to that being the expectation of the professor and the path to best performance (remember that good research is a process, not a one-day lunge), there are the pragmatics of the situation to consider. What if, at the library of your 25,000-student college, the main book in the subject you're researching is "out to another borrower"? What if the older issue of the journal you need is in the "Southern Regional Library Facility," basically a warehouse that it takes two days to get books from? Or what if the first three books you look at have nothing much to say about the exact questions you're trying to answer?

Always start doing the actual searching as soon as you've chosen and understood the topic. Never put this off. It's ridiculous to take a bad grade just because you couldn't *get* *to* the materials you need to study.

TIP #3: Let Your Professor Be Your Guide

Very often the professor will give you a list of "suggested materials to consider"—on the syllabus, on the paper assignment sheet, on the course Web page, on reserve at the library, or sometimes orally as part of the lecture (here's a place where lecture attendance and furious note taking turn out to be super important). In bigger classes at larger universities, the professor will often delegate the dirty work to his or her TAs—asking *them* to make suggestions for research during the section meeting. Don't "underhear" these very important, grade-improving hints. The professor is giving prescriptions of materials that he or she *knows* will work—either because the professor has actually used these sources in his or her own thinking about the issue, or because he or she has seen hundreds of students (well, maybe dozens of students) getting A's in previous incarnations of this course with these very materials.

You'd be amazed how many students don't take professors' suggestions at face value, thinking that somehow bonus points will be given for clever selection of obscure and/or unassigned works. They almost never are. Professors would like to see papers work out. Which is why they suggest to students what to research.

Take the time to track down each and every one of the recommended sources (at least if there's a finite number). Do not second-guess the one who is assigning the research—and the grade.

TIP # 4: Pick the Right *Number* of Sources

Most professors have a pretty good idea of how many sources you're supposed to use in preparing your research paper. Whether they tell you or not. Use not enough and your paper will seem too thin, not well thought-out, or half-baked. Use too many and your paper will seem too schematic, over-argued, and, in the worst case, an attempt to impress the professor simply by piling on sources. Not to mention that in gathering and reading all those sources, you're spending five times the amount of time needed to get the A (which is draining your ability to get good grades in your other courses).

Before starting your trek to the bibliothèque, make a

good effort to determine how many sources the professor expects you to use. A basic rule of thumb is that professors often expect 4 to 6 sources for a shorter research paper (5–7 pages), and 6 to 10 sources for a medium-sized paper (10–12 pages). (For a super-sized paper, 15 pages and above, consult your professor. Expectations here vary widely, depending on the field.) But don't take our word as final on this one. Find out. Consult the paper assignment, together with whatever explanations, hints, noises, or gestures that were made in lecture or in section.

✔+ *EXTRA POINTER*

If your professor has been mum on the issue of number of sources, remember that you have a mouth. Ask the professor straight out. And if your professor won't give any guidance in this area (and all but the most pigheaded—uh, we mean principled—professors will), ask a TA, an advanced student, or a friend who's had the course. Someone's bound to know.

TIP #5: Select the Right *Kind(s)* of Sources

Research for college papers does not involve (at least in the best case) indiscriminate reading of sources. You're not supposed to be making a comprehensive survey of everything that's ever been written on the topic, but a judicious selection of materials appropriate to a college course on the level you're taking.

But how to pick? There's so little time, and so much to choose from.

One very important thing to realize is the difference between *popular* and *scholarly* sources. Popular sources include all materials written for the general public: magazines like *Time* or *Newsweek;* newspapers such as *USA Today,* the *New York Times,* or the *Wall Street Journal;* semischolarly but still popular periodicals, such as *Psychology Today, Scientific American,* or *National Geographic;* finally, more "highbrow" magazines such as the *New Yorker, Atlantic Monthly, National Review,* or *Economist.* In most instances these are not suitable sources for the research paper (though of course there can be exceptions as, for example, if your Contemporary American History paper asks you to consider the reporting of the 2004 presidential election in the popular press).

Also, textbooks are generally not considered suitable fodder for the research paper. Textbooks generally are compilations or syntheses—very boiled down or homogenized versions—of bodies of secondary literature. It's usually expected that you take the research process a step back, by studying the articles, books, or data that themselves are being processed in the textbook. (While no doubt somewhere some professor is assigning a research paper on a textbook, we don't know him or her. And it's a good thing, too.)

What you need in almost every case are *scholarly* or *academic* sources. Books, journal articles, essays, and conference papers that experts in the field have written to report their own observations or studies. In short, *research* materials.

But how are you going to find these? And how are you going to select among these (once you've found them)? For books, the very best place to start is your own library's online catalog (or in rare cases, card catalog). Here you'll be able to search not only titles and authors (if you know them), but also keywords from titles, and subject headings that can be as general or specific as you want. For yak, you can try searching "yak" or "Asian mammal" or "Bos grunniens" or "animals, North Tibet," and so on. Be resourceful in your choice of terms—you might get nothing for one entry, but exactly the books you need for another.

✔⁺ *EXTRA POINTERS*

> Learn how to use Boolean operators—a fancy name for terms such as "and," "or," and "not"—in conducting your searches. What started out as 182 "hits" can quickly be reduced to 10, if you eliminate certain terms that go against what you're looking for. Also, it's good to know about are "wildcard searches"— searches in which you substitute an asterisk or a question mark for some string of letters. These are useful either when a number of words share some letters, or when you're simply not sure of the spelling of some name or other search term. Finally, ace searchers have mastered the art of "limiters"—search boxes that limit the dates of publication to a certain range, the language of the book to one you actually know, and the medium of the publication (print book, e-book, microfilm, or CD) to one you actually have a device to read. Again, fewer results, more relevant hits.

Another good thing to know about—especially if yours is a smaller library or you are writing on a very esoteric paper topic—is the WorldCat database. This is a professional-level but easy-to-use (especially in its basic search mode) database that lists virtually every book in a U.S. library. Ask your reference librarian if your university subscribes (if not, run to the door and X out the word "research" in the sign "University *Research* Library"); also be sure to ask whether you need some special password to access the service. And once you've done a few rounds of searches and perhaps found some book that your library should have but doesn't (World-

Cat will tell you which libraries have what), then go back to see the "interlibrary loan librarian." He or she will be able to order the book for you—no charge, and in many cases electronically (so your material will arrive in plenty of time for you to use in your research).

SOAP *IN OUR HUMBLE OPINION . . .*

The first inclination of most students starting a research paper is often their worst. They log on to a search engine like Google or Yahoo! (or MSN, Look-Smart, or Teoma)—or maybe, for turbo-geeks, some meta–search engine like Dogpile, Vivísimo, SurfWax, or Copernic—and presto, their research is done. Or so they think. The trouble is, no matter how much you refine the terms to be searched, these *general-purpose* search engines often turn up hundreds, or even thousands, of entries. What's worse, the vast majority of results (sometimes even all the results) are not scholarly references, but popular resources that are of little use for the college paper. Look, we are not saying that these search engines are never of use—they are, if you know what sort of thing you are looking for, and/or where you are likely to find it. But if you are just beginning research on a subject, and/or you're not very careful, these all-purpose search engines can lead you where you don't want to go—to popular or nonreliable content that could torpedo your paper. (A good, and comprehensive, report on these search engines can be found at: *www.lib.berkeley.edu/TeachingLib/Guides/Internet/MetaSearch.html.* Have a look if this sort of thing interests you.)

IT HAPPENED ONCE . . .

One time, Lynn assigned students in her Medieval
Art class a paper on the symbolism of a 13th-century
stained-glass window depicting Ezekiel's wheel. The
Old Testament prophet Ezekiel had a vision of four
beasts, each holding a wheel with rims full of eyes. In
medieval art this vision was understood as a prophecy
of the end of time. One student, however, concluded
that the significance of Ezekiel's vision was that it
represented the first sighting of a UFO. When asked
by Lynn how he came up with such an unusual interpretation, the student said that he had simply researched the Internet. It turned out that the Web was
rife with discussions of Ezekiel and UFOs—material
that would be better suited to someone on the day
trip from Mars than a student in a Medieval Art
course!

TIP #6: Use "Gateway" Sources

In every field there are certain works designed to be starting
points for research in that discipline. Perhaps a scholarly encyclopedia that reputable scholars have banded together to
write, like the online *Stanford Encyclopedia of Philosophy*. Or an
online dictionary such as the *Grove Dictionary of Music and Mu-
sicians*. There might be a series of book-length volumes, like
the *Cambridge History of English and American Literature*. And in
some fields you'll find ever-increasing numbers of "Companions" (usually put out by English publishers)—the *Cambridge
Companion to Newton*, or the *Oxford Companion to Indian Theater*,
for example.

If you're lucky enough to have located one of these "gateway" sources (either because it has shown up on your professor's reading list or because you've happened upon it in your own research), be sure to make full use of it. Be aware, first, that the pieces in these volumes are themselves "scholarship"—pieces parceled out to specially selected professors because they have done research on that very limited area or problem. Pay special attention to the way the author frames central issues, and to the "moves" he or she makes in resolving various difficulties. Then be sure to follow out various "leads" the author provides to other potentially productive sources. Sometimes it'll be in the body of the article that the writer points you to opposing—or supporting or confirming—theories, or to other collections of data. Sometimes the bibliography for that piece or for the book as a whole refers you to carefully selected—and up-to-date and authoritative—scholarly treatments of your issue. Whatever is the case in your "gateway" source, a single "hit" can save you about 176 hours of blindly searching the Internet. Seek out such sources, and when you find one (after thanking the gods and/or your professor), use it for all it's worth.

TIP #7: Use Electronic Databases to Find Journals and Articles

In many fields, journals of scholarly articles play an important role in recording and disseminating research. Discoveries are being made every day, and scholars need to get to them in less than the two years it takes to publish a book. So there are quarterly, or monthly, or in some fields, weekly collections of articles published in learned journals. It used to be that libraries subscribed to such journals. No more. Too expensive.

And slow. Now there are electronic journals. And electronic *indexes* of journals, often complete with summaries or abstracts of the various articles contained in those journals.

Many university libraries have "subscriptions" to indexes of journals, and also to "full-text" versions of the articles themselves. They've paid big money for these. *Your* money. So when you think journal literature should make its way into your paper (like when your professor has said, "Be sure to consider some recent articles on the Big Bang theory"), enlist the help of your reference librarian to navigate your way through these shoals.

Every university library worth its salt (and many not worth their NaCl) has elaborate tableaux on its library home page that divide electronic resources into *more general* (though still scholarly) sources (InfoTrac OneFile, LexisNexis Academic, ProQuest, to cite just a few); and more *scholarly* or *"academic"* sources (Academic Search Premier (Ebsco), Expanded Academic ASAP (Gale), JSTOR, Periodical Contents Index, and Web of Science, to name a few that we've had success with). And then there are additional lists of electronic databases by *subject area,* in which, at UCLA for example, you will get 15 databases for "earth and space sciences."

The trouble is, it would take two lifetimes to make good use of these listings if you didn't already know your way around from having done research nine times before in that field. *Get help*. Right away. Go to the reference desk and, once there, try to find a librarian who has experience in the field you're working on. Especially if it's your first time attempting this sort of thing. Many libraries have teams of reference librarians, each with expertise in a particular area or field. Here's a place where tremendous amounts of time can be saved. Picking one right electronic source can save hours of frustration. And pave the path to an A on that research paper.

★★★★☆ **4-Star Tip**

If you decide to "go it alone," you might encounter 100 different entries on the library home page. Talk about information overload. Look for the phrase "electronic resources," and then "electronic resources *by subject*." That will immediately help you narrow your search to, for example, microbiology (rather than natural or hard science in general). And be careful to distinguish "electronic (or article) *databases*," which are lists or directories of different journals, from "electronic (or e-) *journals*," which are the actual journals or periodicals themselves (in electronic rather than print format).

TIP #8: Drive Your Sources (Don't Let Them Drive *You*)

Once they've decided on their sources, the single biggest mistake beginning students make is writing a *report*, not a research paper. They passively assemble the data, then lay it out in an often very comprehensive and elegant way. They cite this source and that, catalog this finding and that, and top it all off with an amazing summary of all that they've found. But they don't have any *organizing principle*— or line of argument or slant on the material—that ties together the various sources and marks the finished product as their own. They are captive to the sources they are studying: "One author says this. I'll write it down. This other author says that. Get it down. Some third author says something else. My paper wouldn't be complete without *it*." The literature is dragging the student through the research rather than the student being the *master* of his or her research.

Any researcher with any talent at all for research will strive to control the materials of study. He or she will constantly be asking: What are the most important points being made here? Why are they significant? What relation do they have to points made by other authors? How well do any of these theories account for the data? Are the arguments or interpretation presented by one or other author cogent or convincing? Has some author made a mistake or drawn an incorrect conclusion? How can all these positions be organized into some coherent and meaningful whole? Can apparent divergences or difficulties be bridged or resolved? These—and a million other questions—are going through the mind of the good researcher. All the time.

And to all these questions (or at least an important subset of them) the researcher will provide an answer. In the form of some thesis, or at least some hypothesis, that is developed throughout his or her thinking about the literature or data. Not his or her "own view" wholly divorced from the materials being studied and the arguments and evidence they contain—but the researcher's own *judgment* or *assessment* about the main strands of argument that emerge from the various materials being studied. The bottom line, the outcome of all that questioning.

Good research is *active, thinking research.* Don't be a sponge, just passively absorbing the data. And good research is research that results in some *conclusion.* Don't be a wuss just looking around at what others have thought.

WANN'AN Ⓐ ?

You'd be amazed how many students get a B+ not because there's anything wrong with their research, but simply because they haven't taken the time—or are

too scared or don't realize the need—to draw any conclusions. The grader is looking, in part, for your own
assessment or evaluation of the materials you've studied. Don't disappoint him or her.

TIP #9: Don't Fixate Too Quickly on What You're Going to Argue

Sometimes students, eager to brand the material as their
own, fix on a hypothesis right away. They read an article or
two, and pretty soon they have it all figured out. And then
they read another source or two and try to "shoehorn"
these additional materials into their first idea of what their
paper was going to be about. Worse yet, students sometimes
distort the views of an author by cherry-picking only points
that support the student's own line of argument. All with
disastrous consequences when the professor knows the literature, and downgrades the student's paper for not having
an elastic enough—and in some cases true enough—thesis
about the materials that have been researched.

*Resist the temptation to decide too quickly what the ultimate
conclusion of your paper will be.* Research (at least good research) is a dynamic and interactive process. You continu-
ally form a *working hypothesis*—that is, a sort of directed
guess—about the material being studied, and then revise
and refine that hypothesis as your thinking progresses.

If research is an exploration of some new territory, it's
not expected (or even desirable) that you know the end of
the road at the beginning. At each stage in the path, the
truth is gradually emerging. Don't commit yourself too
quickly. Unless you're very perceptive (or know a lot about
the issue), your first thought about what you're researching
might not be your best.

TIP #10: Record Your References—and Your Thoughts

Throughout your researching process, be sure to note down whatever sources you consult, and whatever points you encounter that you think will be useful to you in your thinking about the topic. In addition to standard bibliographic information (author, title, date of publication, name of journal), you should always write down the *page number* of any potentially (or even conceivably) important point. You can't imagine how many hours we've spent going back over our research to find some reference to what turned out to be the single most important point in our research effort. Be sure also to record any thoughts and ideas you yourself are having as you work your way through your sources. Students have wanted to commit hara-kiri when they simply couldn't recover the insights they had when first confronting some source.

✓+ EXTRA POINTER

In many courses you hear a lot of talk about the proper style for footnoting quotes or ideas taken from others. APA, MLA, CBE, APSA, AIP. It's alphabet soup out there. And who's Turabian, and what does *The Chicago Manual of Style* (15th edition) have to do with footnoting, anyway? Figure out what style your professor is expecting *before* taking your research notes, so you'll have all your references properly prepared. And if your professor hasn't come down on any side of the footnote wars, pick one style you like and use it consistently in your paper. You might enjoy navigating your way to any of the following Web sites, which will educate you on the minutiae of footnote styles:

www.lib.duke.edu/libguide/cite/works_cited.htm

memorial.library.wisc.edu/citing.htm
   (No www. in this one.)

www.columbia.edu/cu/cup/cgos/idx_basic.html
(This last one includes information about how to
   cite a broad variety of Internet sources.)

TIP #11: Distill Your Research into an Outline

Reading notes are one thing. A plan of action is another. Your researching process should always culminate in an outline. Some reconstruction, in some organized fashion, of the research you've done. And some pointing forward to the writing you're going to do. Try to figure out what the *structure* of what you're going to write will be. What are the central points, what are the supporting points, and what points can be sent back to where they came from? Sure, you'll have additional thoughts as you actually write your perfect paper. But having a pretty good idea of what you're going to argue already at the end of your researching and "thinking out" stage is the single best guarantee of an A paper. *What you write will be what you think.*

As always, we've saved our best—and by now most anticipated—tip for last:

BONUS TIP : Make a Trip to See the Man (or Woman)

If by the end of your researching process you're still not 100 percent—or at least 99 and 44/100 percent—sure about what you're going to argue in your research paper (and how you're going to argue it) . . . (You know the rest. Which doesn't make it less true.)

Review Session

The research paper can be either one of the most daunting, or one of the most valuable (indeed exhilarating) experiences of one's college career. It helps if you know what to do. Here are our top 11 tips for doing the research:

1. **Get out of your head.** Go to where the object of your study lives (Tibet, the library, wherever).

2. **Start your research ASAP.** Why run out of time, or get caught short when the book you're looking for is out to another borrower?

3. **Let your professor be your guide.** Track down each one of his or her research "suggestions." They're guaranteed to work.

4. **Pick the right number of sources.** Not too many. Not too few.

5. **Select the right kind(s) of sources.** Scholarly. Not written by some 14-year-old kid from Brooklyn in his basement.

6. **Use "gateway" sources.** Scholarly encyclopedias, dictionaries, and companions that point beyond themselves to (a finite number of) useful other sources.

7. Use electronic databases. The home of articles.

8. Drive your sources. You're the master of your research.

9. Don't fixate too quickly on what you're going to argue. He or she who lunges is lost.

10. Record your references—and your thoughts. The best time-saver known to man.

11. Finish it off with an outline. Your mirror back. And your map forward.

Bonus Tip. Go see the professor (or TA). He or she will be especially sympathetic and willing to help when you've done so much of the legwork already.

CHAPTER TWELVE

Do's and Don't's for Going to See the Professor

As your work on the paper progresses, many roads lead to the professor. Lots of students at some point in their thinking about the paper (be it an analytical or a research paper) find themselves confused about one or another point, or unsure whether they're on the right track. If and when this happens, the simple solution is to head directly to your professor's (or TA's) office hour to consult. Your little tête-à-tête with the professor might last only 15 or 20 minutes. But this quarter of an hour could have more impact on your grade than all those hours spent holed up in the library—times two. Like any other human interaction, the office meeting with the professor can hit the rocks. Especially if you don't have the right combination of academic and interpersonal savvy to pull it off. Follow our do's and don't's and you'll be sure to reap the full benefit of your excursion to see the one who holds your fate in their hands. Your grade fate, we mean.

DON'T : Don't Be Afraid to Go See the Professor

Some students avoid going to see their professor simply because they're intimidated. Perhaps they have heard how erudite the professor is, how much of an expert in the field he or she is, or how much he or she has published. And the last thing they want to do is have a one-on-one discussion with the world's leading authority on the topic that they have to write about in their paper.

But if you have questions about your paper—if you really don't know what to argue in the paper, whether a given point should be included or left out, or what point should come next in your presentation—then you should just suck it in and go. Don't worry, it won't be as scary as you think. Did you know that this professor is just a regular person once he or she steps out of the academic world? That there are normal, everyday people who flip this guy (or gal) the bird for cutting them off on the freeway? You have a question that needs to be answered. And your professor is clearly capable of giving you an answer—one that could really help your grade. Live brave. And get your answer. Who knows, a good dose of adulation—not to mention genuine intellectual engagement—might do your professor some good.

You see, some schools have an environment in which professors are seen as there to help. But at other colleges, professors are held up as demigods. We have a friend who taught at Harvard for a while and was astonished to have students coming in quaking with fear. Once he moved to another, somewhat less prestigious school, in Southern California, he was amazed to find students perfectly comfortable treating him in their standard snarky way. Same super-nice and supersmart guy, completely different student reaction. When it comes to seeing the professor, try to forget about the school's culture and the professor's reputation.

Another thing to keep in mind is that you have prepaid for the course, and that part of the piles of money you (and your classmates) have paid goes to pay for the prof's office hour. The one you're weighing whether to exploit. Would you go out and buy a pair of Manolo Blahnik shoes for $500 and then be too scared to wear them? Would you buy a Harley and be too timid to ride it? Why pay good money to study at a school with distinguished faculty and be unwilling to ask them to do what they're getting paid to do?

DON'T : Don't Worry about Seeming Stupid or Hurting Your Grade

Some students are loath to go to see their professor because they think that the questions they have will seem too basic or plain stupid. Or because they find the paper—and perhaps even the whole course—to be very hard. They think, "If I go meet with the professor, I might make a terrible mistake, one that could really hurt my grade. After all, the same professor who's answering my dumb-ass question about what to put in my paper is the one who's going to be grading that same paper—in a little over a week."

But there's really no problem here. There's no need for conservatism. Office hours are not graded activities. Not even as part of the class-participation grade (if even there is any). Think about it. Which will make a better impression on your professor: if you come in, make a mistake and it's corrected, or if you write that mistake into your paper in bold, 12-point, Times New Roman font, and it's corrected— this time with points taken off? When it comes to papers, talk is free; writing is where the jeopardy comes in. So don't worry about talking stupid, just worry about *writing* stupid.

And don't let the fear of making mistakes keep you from seeing your professor. The office hour is a free throw. If you go in having given any thought at all to the assignment—and if you treat the professor with even a modicum of politeness (always a swift idea)—you have nothing to lose and everything to gain.

It's particularly important to capitalize on this advantage in killer courses. Most professors respond particularly well to students who find the course tough but are bearing down and working hard. In some instances that's even part of the "plot of the course" in the professor's mind—how he or she thinks it's *supposed* to go. Indeed, professors often find themselves secretly rooting for students of this kind, wanting them to do well. And next thing you know, the professor is giving that student extra help, even more than that professor might ordinarily give. It's well worth making a mistake or two to get this kind of deal.

How to Make a Good First Impression on Your Professor

Be a pleasure to meet with. Make the professor happy you've come in. And leave the prof looking forward to your return visit.

☑ **Go to scheduled office hours (if possible).** Though the professor might be willing to see you "by appointment," such visits are extra work for the professor and can mean wasted time during the regular office hours.

☑ **Ask nicely for any special appointment.** Ask the professor when it is convenient for him or her to meet you, rather than reciting your almost full booked schedule. And don't be a "no-show." A no-show student equals very ticked-off professor.

☑ **Be on time, and bring your own notebook and writing implement.** Show that you came ready to work and don't need to bum pens or paper off the professor.

☑ **Smile when you come into the office.** It's nice to be nice, and besides, the nicer you are, the more likely it is that the professor will invest time in your case.

☑ **Never talk while your professor is talking.** Can you spell R-E-S-P-E-C-T?

☑ **Be sure to thank the professor cordially when you're finished.** Not one in a hundred students does it, but face it, the professor is human, too—and will remember these little niceties when you're back in a couple of days for more help.

DON'T : Don't Expect the Professor to Give You the Answer

Some students make the mistake of thinking that they can just go to the professor's office hour and the professor will give them the answer to the paper assignment. But see, if professors wanted to "give you the answer," they wouldn't bother giving you the paper assignment. They'd just write the paper for you. Or skip it altogether. No, professors aren't the "Answer Man (or Woman)"—though, on some level, every student wishes they were. Professors think it's *your* job to come up with the ideas, to have some thoughts about how to develop them, and to have at least a preliminary guess as to where the paper will wind up. Professors will, however, help you get your paper started, they will help you move it along, they will help you navigate it around the bends, and they will help you assess what you've really shown in the paper.

When you deal with professors, keep in mind that if you put nothing in, you'll get nothing out. If the professor senses you haven't put any energy into the paper, he or she is unlikely to give you much time or much feedback. But the more professors see that you've been wrestling with the issues posed by the paper assignment, the more likely it is that they'll give you information, suggestions, and advice. *Whenever you go to see a professor, have a few specific ideas or questions prepared.* Show what work you've done and ask particular questions about the trouble you're having. If you've got nothing like that to ask, you need to do more legwork before embarking on that pilgrimage to the professor.

DON'T : It's Good to Come Prepared, but Don't Ramrod through an Agenda

You are not likely to have a fully successful meeting with a professor if you come in with a point-by-point agenda that you are determined to get through, no matter what. Here's how such a meeting typically goes. You start explaining your points, one by one. The professor sits and listens for a while, then when something strikes him or her as interesting, suggestive, insightful, or significant, he or she stops you and raises a question or thought. Often quite briefly and tentatively, at first. You provide some brief answer (perhaps even meeting head-on the question asked), but then you put your head down and move on to the next point you had prepared. And then the next one, and the next. These later points are sometimes less important or even irrelevant. At least from the professor's point of view (which, when it comes to grading, is the only point of view that really counts).

Keep in mind that the more a professor engages in a topic, the more likely it is that he or she thinks it's important for that paper. So don't shut down the discussion of a central issue in order to jam through your agenda. Rather, run with the ball and see where it goes. Look, there's nothing wrong with having an agenda (though a more fluid series of "working points" is often better). But be sensitive to the other half of the interaction. The professor who's trying his or her level best to guide you in the direction of the right answer. And the good grade.

We've witnessed many an office hour in which a student's preconceived agenda put an early end to what could otherwise have been a very productive discussion. Sometimes the student simply wants to show off the good work he or she has already done. Sometimes the student is inattentive or insensitive to clues the professor is giving about what direction the discussion should take. And sometimes the student is simply unwilling to participate in the normal give-and-take of a cooperative (and dynamic) exploration of a topic or issue. But whatever the reason, the student has deprived him or herself of a great opportunity. To engage one-on-one with a genuine expert in some field, and to have the student's own ideas worked on by the professor.

SOAP IN OUR HUMBLE OPINION . . .

In all your meetings with the professor, we suggest you *take notes* on the issues that are being discussed. This way, whatever you learned from the meeting you won't forget. And it'll keep you 100 percent focused on what the professor is suggesting. Though it might seem a bit awkward to take notes in the office hour, most professors won't mind note taking a bit. They might even think it shows motivation, a desire to learn, and a sense that the student values what the professor is saying.

Once in a while it happens. You're sitting in the office with a professor and you have a true meeting of minds. Genuine intellectual engagement. The professor is talking about some subject—perhaps his or her own research—and all of a sudden you could see yourself actually going on to work in that field. Or you're putting up some idea, and the professor sees in you a version of him or herself at that stage of intellectual development, and goes on to encourage you to pursue your interests in that area. Though it is rare, when this happens it is a genuinely life-changing moment. Stop taking notes. Listen to what the professor has to say. What started out as a discussion of some or other paper topic has morphed into something much more powerful and important. Something that will alter your life.

If this ever happens to you, we'd like to hear about it. Take a moment to e-mail us at: **Lynn@ProfessorsGuide .com** or **Jeremy@ProfessorsGuide.com.** A place of honor is reserved for you in the next edition of *Professors' Guide.*

DO : Be Willing to *Rethink* Your Ideas

Imagine this scenario. A student comes in to discuss how best to write the conclusion of his or her paper. But the professor quickly realizes that the student has made a pretty big mistake in the main body of the paper. And that the conclusion is the least of the problems here. Obviously not so pretty.

Professors don't relish encountering a student who has

gotten well into the paper, but is substantially off track. A problem that's especially acute when the student is hoping the professor will help him or her just shut it down—while the professor knows that what's really needed is a trip back to some earlier stage, perhaps even step one. It doesn't take a Ph.D. to realize that if you tell students straight out that all their hard work to date is for naught they are likely to get pretty upset.

Many professors confronted with this situation resort to gentle hints. They try to suggest that there are "problems" with earlier stages, that some previous point needs "rethinking" or "recasting." They point to some earlier stage in the outline or draft you've been kind enough to provide them with, and ask a question or two about why this or that is there. They resolutely avoid talking about any issues that come up *past* the point where the paper has gone off the track. But if the student fails to take the hint, the professor isn't about to hit him or her over the head with a baseball bat. Be attentive to clues that some earlier stage in your argument is defective. Don't let your eagerness to get this thing out of the way blind you to the need to redo or rethink parts of the paper that you thought were finished.

And it's often a good idea to go to see the professor sooner rather than later. When the ideas are just beginning to percolate, and you don't have so much invested that you can't change your ideas in light of the professor's input. *Involve your professor in the generation and development of the ideas.* While there's still plenty of time.

DO : Be Willing to *Extend* Your Ideas

There may be times when you're having a meeting with the professor that is going fabulously. The professor is really liking your ideas and telling you how insightful they are. But then your professor drops a bombshell. He or she suggests you might want to do some additional work! Read another text, look up another source, get an article from Interlibrary Loan, consider an additional ramification of your idea, answer an objection, provide another example to support your point, whatever. And you're left wondering whether you really need to do the extra work, given that the professor seemed to like what you'd already done just fine.

No, you don't *have* to. Unless you want an A. You see, professors really enjoy seeing students who have already made good progress on their papers. And when professors see that, there's nothing they enjoy more than offering up words of praise and making suggestions for even more ways to make the paper better. If you get these sorts of suggestions, it's a pretty good idea to pursue them. Sure, it means more work. And yeah, it requires you to push harder when you've already pushed plenty (and done a pretty good job at it, too). But it is exactly this professor-directed extra work that could land you an A– instead of a B or B+. Or an A instead of a B+ or A–. (And, of course, there's the satisfaction to be had at a job well done. Really well done.)

DO : Make Positive Use of Criticism

It's a very real possibility that any meeting you have with the professor will involve some element of criticism. Of you, in particular. Especially if you give the professor a full presentation of your ideas or even an outline or a draft of your project. It's always upsetting to get criticism. That's one of the hardest parts of college. And of life. You've invested your heart and soul (or at least a few hours of your brain time) in your work, and now your professor is coming at your work like a freight train. With no sign of slowing down.

But it's an intrinsic part of the learning process—and of the grading process, too—that professors raise criticisms about student work. Not necessarily in a nasty way (at least, most professors aren't deliberately *trying* to be nasty), but often in a reasonably direct way. Professors have to be able to communicate what the deficiencies of a paper are, if students are ever to correct their thinking and, in so doing, improve their grades. Whether they like it or not, college students need to learn to live with criticism. The strongest students are the ones who can turn criticism to their advantage. The ones who respond to the criticism in ways that can enhance, rather than detract, from learning. Which all boils down to: Don't get upset, get busy. View criticism as an opportunity, not a problem. If a professor will give you a full explanation of problems in your outline or draft, he or she is giving you a real shot at improving your paper—and your grade!

You Can Do It!

It's helpful to remember that some of the upset you feel when your ideas are criticized is a matter of perspective and timing. During the meeting, when the professor is evaluating your ideas, it's natural to feel that you're under attack, being assaulted from all sides. But later on, when your paper is handed back with a crowning A, you might feel that the professor's questions, comments, and objections were on the mark. And that they turned out to be extremely helpful in getting you that A. Resist the temptation to be overwhelmed by the emotions of the (first) moment.

✓+ *EXTRA POINTER*

> Sometimes, when all seems lost, it's worthwhile ask-
> ing the professor what he or she would recommend at
> this point. If every word that's come out of the profes-
> sor's mouth is critical or disparaging, a radical break
> may be in order. Simply say, "This doesn't seem to be
> going too well. What would *you* recommend I do?"
> Then don't talk, and listen very carefully. Sometimes
> the professor, so struck by your candor and willing-
> ness to come for help, will actually begin to make
> useful suggestions. Press for more. Ask, "What would
> *you* do next?" Then listen more.

DO : Tell the Prof When You're Not Understanding

There is no point bluffing your way through a meeting with
your professor. No point pretending that you understand
what he or she is saying when you don't.

How is your professor going to help you if you're not
honest with him or her? Let your professor know if you
can't understand something he or she tells you. And don't
worry, your professor can handle it. It's not exactly like it's
the first time the professor caught wind that some student
didn't understand something. One of the biggest impediments to student learning is failure of communication between professor and student. All professors want their
students to learn the material they are teaching. Whether
they succeed or not depends largely on good two-way
communication. It doesn't help the communication any if
a student feigns an understanding that he or she doesn't
have.

Most professors can simplify their presentation—down a

number of levels, if necessary—and are happy to do so if a student needs it. And if the professor is going too fast, ask him or her to slow down. They can talk slower, too. It's easy for a professor to get caught up in the excitement of the moment and fail to notice that the student is lagging about four steps behind.

DON'T : Never Lock Horns with the Professor

The worst meetings between professors and students are the ones that turn into major-league fights. Usually this gets started when the professor says something that the student feels is dismissive of his or her work. In response, the student doesn't just engage in a normal intellectual discussion— he or she becomes very angry, sometimes to the point of yelling or even swearing at the professor. Yes, it happens. And it's not a good thing for either the student or the professor. When you go to see your professor, it's best (everything else being equal) not to engage your professor in mano-a-mano combat—even if he or she says things about your work that you don't want to hear.

If despite your best intentions—and our fine advice—you still end up in a pitched battle, it's not likely that the professor will take it out on your grade. Most professors are committed to basic fairness. The professor will probably grade your paper without regard to the outburst and average it in with the rest of your work (just as he or she does for all the other students). Or maybe the professor's not even your grader (in which case you blew your stack for nothing).

Still, it'd be a nice touch if you went back to apologize during the next office hour. Professors like that sort of stuff. Who knows, maybe the professor is feeling bad, too. While you're there, you might do something to really help your

Jeremy Remembers When . . .

One time when I was a TA at UCLA, my office-mate Ron was talking to a particularly contentious student. I could overhear the discussion starting to boil over, and the next thing I saw, the student was clenching his fist on the desk. Without missing a beat, Ron jumped to his feet and said, "Mr. M., if you don't unlock your fist, you're going to find your head on the other side of the room." And he meant it. (The student, suitably chastised and maybe even a little scared, seemed very polite after that.)

grade. Like having a *nondefensive* discussion of what you're going to argue in your paper.

DON'T : Don't Come Off as Mr./Ms. Know-It-All

There are some students who decide that *they're* the real experts in the field. That they have an incredibly brilliant idea for their paper. That they are able to solve a problem that no one has been able to solve before. So any suggestions or criticisms that the professor raises are either ignored or taken as signs that the professor is simply too stupid to understand.

It goes without saying that professors do not like meeting with know-it-alls. But more important, the boy-genius or girl-genius attitude is not very helpful for the learning process.

Even if you think you've hit upon a major discovery in writing your paper, never come in to see the professor with the attitude that you are absolutely, positively, 100 percent correct. And that you don't need to pay attention to any corrections or suggestions for improvement. Even the leading authorities in a field have to consider objections to their theories or take into account new evidence that comes up. That's what learning and intellectual progress are all about. And face it, it's not really so likely that you, as a beginner in a field, are going to have come up with ideas that will land you a MacArthur Fellowship (at least not without some further tweaking).

DON'T : Never Tell the Professor What You Really Think of the Course

Always remember that the office hour is not the time to level with the professor—if by "level" you mean telling the professor that you hate the course. That you wouldn't be taking it except that it's required. Nor is the office hour a good time to ask the professor when the course will get better or the lectures more interesting.

No one wants to hear *that*. And no one will give you a good hearing after you've kicked him (or her) in the face.

DON'T : Don't Second-Guess the Results of Your Meeting

Sometimes a student will bring some ideas to the meeting with the professor, and the professor will suggest changes to, or improvements on, the ideas. Everything seems to be going great. Everyone walks away from the meeting with that feeling of accomplishment you get only after making some real intellectual progress. But then, when the student sits down at the computer to write, something strange happens. The student decides to revert to plan A. To write what he or she had planned *before* the meeting with the professor.

You should never second-guess the professor's suggestions—or the additional thoughts you yourself had during the meeting—only to go back to your old way of thinking about the paper. If the old way was so great, the professor wouldn't have suggested an alternate route, and you wouldn't have had the new thoughts that you did. And why did you even bother to go talk to the professor if you were just going to do what you wanted all along? Students who backslide are giving back all the extra points they would have gotten if they had followed the advice that their professors so nicely gave them. Not to mention losing the intellectual benefit of a productive work session with a professor on a topic of joint interest.

SOAP IN OUR HUMBLE OPINION . . .

Some students, in large courses with lots of TAs, confuse going to see the TA with a trip to the mall. They go from store to store, and keep coming back, till they find the merchandise they want. Or maybe they don't buy at all. Don't succumb to this mind-set. TA-shopping can be highly inefficient (you waste time waiting till that other TA has an office hour). It can get you very confused very quickly (as you try to process all sorts of different, and sometimes even contradictory, advice about how best to proceed in your paper). It can represent avoidance-behavior with regard to the real task at hand (getting down to writing some words on that blank piece of paper). And far and away the worst, it could cause you to ignore specific suggestions from the person whose opinion counts most (your own TA who in just three days is going to be grading your paper).

Of course, there can be cases in which it is beneficial to seek a second opinion. Sometimes another TA (or the professor) can suggest ways in which you can probe a point more fully, or extend your thinking more broadly, or consider other sources than your own TA has suggested. Sometimes your TA simply didn't have the time to address your concerns fully enough. And sometimes the communication between you and your TA just isn't that good. But whoever else you see, always keep in mind that in the end it's going to be your own TA who (unless otherwise indicated) will be deciding on that grade at the end of your prize paper. You won't be able to duck him or her, so be sure you adequately address whatever he or she has said.

DO : Go Back to the Professor If You Have More Questions

After you've seen the professor, you might find yourself with new questions that you'd like to ask. If so, you might wonder if it's okay to go back for another meeting. Or whether, if you do, you'll be regarded as a total pest.

You are probably in the best position to read the vibes of your professor. But in general, most professors don't mind if a student comes in a couple of times while preparing the paper, especially if the first meeting went well. You could probably even come in a third time if needed. Just watch for clues that the professor has had enough of you—like if he or she slams the office door, or hangs out a Do Not Disturb sign, the minute he or she spots you coming down the hall.

How to Go See Your Professor—
Electronically

Though you might not have thought about it, most professors are willing to offer help and answer questions by e-mail. Current trends suggest that e-mail will become an increasingly important mode of communication between student and teacher—and an increasingly useful way to enhance learning—as time goes on.

E-mail communication with professors usually works best when your questions are relatively confined and can be answered in a paragraph or two. Longer questions or questions requiring significant interpretation and judgment are almost always best taken up in an office hour.

To make the best use of the e-mail option when preparing your paper, follow our top 10 tips for e-mailing your professor:

1. **Be sure that you send your e-mail to an active e-mail address**—one that your professor actually uses (not an address assigned by the university that the professor hasn't even heard of).

2. **If possible, send your e-mail from a reputable university account,** not one whose address could be confused with an adult or weirdo site (professors are less likely to respond to mail from Hotchick or even Squidboy).

3. **Write an informative—but nondemanding— subject line.** Professors may be getting as many as fifty e-mails a day, and you're likely to get a better response with "request for meeting" than with "must meet—NOW!"

4. **Be sure you begin your e-mail with a pleasant and respectful salutation** ("Dear Professor Hyman" would be a better choice than "Yo, Jero!"); and end with a gracious "thank you for your kind help" (especially if the professor actually provided, or could conceivably provide, kind help).

5. **Don't expect your professor to give you instant feedback** (professors don't like to feel like they are on call 24/7; they're Ph.D.'s not M.D.'s).

6. **Don't make unreasonable demands on your professor** (like asking him or her to e-mail you a copy of the paper assignment that you lost, and while they're at it, include their notes for the classes that you missed).

7. **Don't submit drafts of papers** for the professor to review without confirming in advance that he or she is willing to review drafts (amazingly enough, some professors just aren't up for reading 50 pages of drafts the day before 250 pages of papers are submitted to them). And don't submit your completed paper electronically unless your professor specifically allows you to (downloading and printing 250 pages of papers might also not be your professor's cup of tea).

8. **Confine your e-mails to academic issues** that are directly related to the paper assignment or the

course (don't ask your professor to suggest places
where you might locate a good printer, now that
your printer just broke down).

9. **Don't e-mail your professor incessantly** (keep
in mind that your professor could have hundreds
of students and doesn't want to see your name in
his or her inbox on a constant basis).

10. **If your professor doesn't answer your e-mail,
don't assume that he or she doesn't care about
you or hates you** (the professor might not even
have gotten the e-mail in the first place, or might
have mistakenly deleted it—most don't have the
best hand-eye coordination in the world). Either
retransmit your message (with an apology for double posting) or get your question answered the
old-fashioned way—in the office hour.

And finally, and most important,

BONUS TIP: **Keep in mind your prof is a fellow
human being,** and might actually read (and have feelings about) the e-mail you've dashed off.

Review Session

There are as many ways for an office hour to go wrong as there are kinds of professors and kinds of students. Like any human interaction, the success of the encounter depends on dozens of intangible factors—the personalities of the two people involved; how those personalities engage (or fail to engage); how certain key comments are taken (or not taken); whether the people establish a rapport (or don't feel all that comfortable together); how relaxed the meeting is (or how much of a rush the people are in); and even what each had for breakfast (or didn't).

Nevertheless, there are some objective strategies you can use to maximize your chances of a productive meeting—and to improve your grade on that all-important paper. Here are our best ideas grouped under four headings (it'll be easier to remember them that way):

1. **Don't be afraid to go see the professor,** or worried that you'll seem stupid or damage your grade. The professor (or TA) is just an ordinary Joe or Jane, he or she knows that the course is difficult for some students, and besides, office hours don't even count for the grade.

2. **View your meeting as a joint exploration of the topic.** Realize that the professor won't just give you the answer; that you won't be able to ramrod through an agenda; that you might have to rethink, or extend, the ideas you came in with; and that you might have to accept criticism or tell your professor when you're not understanding something. (But in exchange for these minor unpleasantnesses you might get one of the best experiences of your college career—and in any case, the best help for an A that money can buy.)

3. **Don't act like a jackass (even accidentally).** No professor likes students who are combative, or conceited, or overly "honest." Always try to imagine the other side of the interaction (that is, how *you* might be seeming to the professor). Then act accordingly.

4. **Keep your momentum going after the meeting.** Try to incorporate the results of the meeting into your own thinking. Seek another opinion if you think it would be helpful. And feel free to make a return visit to the prof if you've got more to ask.

The office hour is one of the best—and one of the most underused—resources of the American University. Learn to use it well.

CHAPTER THIRTEEN

Top 10 Tips for Constructing the Perfect Paper

If you've followed our method so far, you're well on your way to writing the perfect paper. You've understood exactly what the paper assignment is asking and what kind of paper you're supposed to be producing. You've done the analysis and/or the research, you've made a trip (or two) to the professor and incorporated his or her comments into your thinking, and now the moment of truth has arrived. It's time to polish up your work and take your final steps toward that A paper. Time to make sure that the product you create—the paper that the professor is going to spend about 15 minutes reading through—is a worthy reflection of all the effort you've invested. Don't cut a corner here. At the grade junction. Have a look at our top 10 tips for constructing the A+ paper.

TIP # 1: Get Right to the Point

A great paper needs to get off to a great start. Don't plan to ease into the paper topic with a long introduction (unless otherwise instructed). The professor doesn't want to hear general background information, or how the assigned topic is interesting but difficult, or what thought processes you went through while writing the paper. Professors don't want to suffer through elaborate stage setting, whether it takes the form of flowery opening paragraphs or ridiculously basic ones.

Professors want to see your answer ASAP. Like the hungry diner, they want to get to the meat and potatoes right away. So the introduction to your paper needs to be short. It needs to tell the reader what you are going to do and how you are going to do it. It needs to include only the materials relevant to the task at hand. *Begin to answer the question in the very first sentence.*

IT HAPPENED ONCE . . .

A professor and a TA at the University of Pennsylvania once conducted a grading experiment. They each graded the same set of papers, but the TA read the whole paper while the professor read only the first paragraph. Amazingly enough, the grades the two graders gave for each paper came out virtually the same. Apparently you *can* accurately judge a book by its cover, or at least by its first page.

Keep in mind that when professors are reading papers, they are impatient. They have a big stack to read—sometimes 50 or 60 or more. When they get to yours, they have one question foremost in their mind: What grade am I going to give this one? The minute they start reading, they are looking to see what the content of your paper will be, and how this content stacks up against their expectations. That's why it's important that your answer to the question posed by the paper assignment come up quickly. If by the end of the first page professors don't see any specifics on what your paper is going to say, they will most likely assume that you don't know the right answer.

TIP #2: Advance a Thesis

Every college paper needs to have a thesis—and one that can be expressed in a single clear sentence. Think of your thesis as the mission statement for your paper. It's the one-sentence answer to the issue raised or the question asked by the paper assignment. It's the way you would fill in this blank: "What I show in this paper is that _____."

COLLEGE SPEAK

A thesis is "a proposition laid down or stated, especially as a theme to be discussed and proved or to be maintained against attack" (*Oxford English Dictionary*, p. 3288).

It's extremely important that you think very carefully both about what your thesis is and how you will express it. Your whole paper will essentially depend on the strength of your thesis. Be sure to review any information you might have gleaned from meeting with the professor during the office hour. And look over your outline from either the analysis or the research you did previously to see what thesis best captures the strongest point of your work—and the best answer to the question asked. Once you've established a solid thesis, you've laid out the goal, and set the agenda, for your entire paper. This will make it much easier to finalize what you need to be doing in the body of the paper.

The thesis gives your paper a point, and college professors expect papers to have a point. A paper without a point is a *report*. And a bad one at that. The point of your paper should not be something that just emerges from what you say. It shouldn't be something the professor has to search

Lynn Remembers When . . .

One time when I was teaching at Vanderbilt University, I wanted to ensure that all the students in the class were clear about the need for a thesis in their papers. So I announced to the students that I would absolutely, positively be expecting to find a statement of their thesis at the beginning of their papers. If they didn't know what their thesis should be, I said, students shouldn't start writing but should come see me in my office. The next afternoon, I noticed a line of students going up the stairs, starting at the first floor and ending on the third floor, right at my office door. I had to spend the entire afternoon helping students formulate their theses. Imagine what would have happened if I had given instructions for the whole paper!

around for, or figure out, on his or her own. It is something that has to be explicitly declared at the outset. And the thesis is where you declare it. When professors read papers, they are immediately on the lookout for what the thesis of the paper is and how good it is. What they find will make a big impact on their determination of the grade.

TIP #3: Be Clear Throughout

One of the hallmarks of a good paper is that it is clear throughout. At every point, from beginning to end, the reader is able to understand what your main ideas are and what support you are offering. A less successful paper, on

Professors' Perspective

Most professors who assign papers experience a time when a student comes in to see them after being told that his or her paper was unclear, vague, hard to follow, or confused. The student points to specific places in the paper where he or she made certain points, then goes on to explain the answer to the question in good detail. And then the professor understands exactly what the student meant and realizes that the student really did know the answer all along. The trouble is, the professor has to grade what's on the paper. And can't grade what's in the student's head. Or how well the student can explain what he or she wrote, after the fact.

the other hand, is often muddled, or vague, or confused. The reader is left scratching his or her head, having a hard time figuring out just what you mean and how you have argued for your main claims.

Any professor who has taught for any length of time is primed to spot sentences, indeed sometimes even whole ideas, that are expressed confusedly or with lack of precision. Our friend David, a professor at the University of Vermont, once came across this sentence in a student's paper: "While lines of fit seems to undermine the exaction of accurate science, in actuality the process is an admittedly flawed and conveniently resistant method." It's pretty easy to say what's gone wrong here. The writing is confused or ungrammatical, it's not at all clear what (if anything) the student is talking about, and one of the words used ("exaction") doesn't mean what the student thinks it does (if indeed it's a word at all).

There is no recipe or set of rules for producing the paper that is clear throughout. In the end, any paper that produces a clear understanding in the mind of the reader of what you're trying to say *just is* a clear paper. Nevertheless there are some techniques that you can use to raise the level of clarity—and with it the level of your grade—of any paper.

10 Ways to Improve the Clarity of a Paper

If you sense that your paper isn't as clear as it could be, try doing one or more of these things:

1. Explain a point more fully or in a different way.
2. Separate out two points by drawing a distinction or by treating them one at a time.
3. Make explicit an assumption you are making or a point that you are presupposing.
4. Give an example or two to illustrate or clarify a point.
5. Define a key term, especially if it's being used in some special or nonstandard meaning.
6. Take out some content that is detracting from the main point or is confusing the issue.
7. Take out any points that are repetitive. Say each thing once and say it well.
8. Make sure your paper moves forward. Take out any sentences that look back unnecessarily, or that reintroduce points for no apparent reason.
9. Make sure it's clear what each pronoun ("it" or "they") is referring to.

10. Correct any errors of grammar, spelling, or usage. Even if the professor doesn't penalize for poor writing mechanics (most don't), a badly written paper suggests unclear thinking and simple lack of caring (things that the professor will surely take off for).

TIP #4: Write for a Reasonably Intelligent Person— Not the Professor

Anytime you sit down to write, you are choosing an audience—even if you aren't aware you are doing so. It's natural enough for college students to assume that the audience for their paper is the course professor. After all, the professor (or the TA) is the person—in some cases the *only* person—who is going to be reading the paper. And he or she is the only one who is going to be assigning the grade. What better strategy could there be than to select the professor as your intended audience, to write in such a way, and on such a level, that only he or she will understand?

Don't write for the professor. Instead, write for a reasonably intelligent person, but one who hasn't taken this course and isn't already familiar with its content. For *this* reader you will need to explain in detail all the points that he or she doesn't understand; you'll need to define any terms that he or she wouldn't understand; and, most important, you'll need to explain completely enough that he or she understands what you are trying to say. It's not enough to gesture at points, to mention them in a sentence or two. You have to provide a complete enough explanation so that an untrained, though smart enough, person can grasp your answer. Really get his or her mind around your idea(s).

You don't want to write for the professor because when you do that, you start by assuming an audience who already knows all the material. Why should you bother to explain stuff the professor already knows like the back of his or her hand? That sounds like a total waste of time and energy.

But to get a good grade, your paper needs to *show* the professor how well you understand the material. Your paper can't rely on what the professor knows. The professor knows that *he* or *she* knows the material; the professor is trying to determine whether *you* know it. The way you demonstrate this is by fully explaining the concepts and terms so that you don't have to be a professor to figure out what you're saying.

We used to tell students that they should write their papers with their roommate or parents as the intended audience. This got us into trouble. We would get extremely simplistic papers that students said were written for their blockhead, hungover roommate. And we would get very terse and difficult-to-grasp papers written with astrophysicist moms in mind. When thinking of your audience, try to find a middle ground. And when in doubt, *err in the direction of more explanation, not less.*

TIP #5: Use Simple and Direct Language

Students sometimes think that by using lots of big, unusual, "college" words, they will impress the professor. They look over lists of fancy words, use a thesaurus, or pore over their class notes to find special technical terms to insert into their papers.

This rarely works. What the professor is really looking

for is simple, clear, and direct language. Sure, if there are technical terms that bear on the issue, the professor wants you to use them and use them properly. But the professor doesn't enjoy it when students artificially bring in terms and words just for the sake of sounding smart.

That's because when professors read papers, they are making a directed search for specific points. Though there is often no "right" or "wrong" answer to a paper assignment, professors usually expect to find certain particular points addressed in the paper. Often in grading, professors will put a check mark or make some notation in the margin when they come to one of these key points. If you write in simple and direct language, the professor will be able to more easily locate what he or she is looking for and credit this toward the grade.

Also, when you write in simple and direct language your paper has more force. Your arguments become more effective. You are not just "dancing around" a point, you are asserting it, simply and directly. Using direct language also helps you avoid misuse of terms and words. This can often happen when students are using words that they don't understand, just to look good. This strategy can end up amusing your professor. But it will do nothing to get you a good grade from the professor (no matter how funny a mood he or she is in).

✓⁺ **EXTRA POINTER**

Is it a good idea to include lots of quotes in your paper? Is it a good idea to let the experts do the talking for you? Actually, no. The professor is interested in what you, not your sources, however eminent, have to say. So be sure to:

☑ Use quotes sparingly. Establish for yourself a "quote quota."

☑ Quote only materials that directly relate to, and advance, your arguments.

☑ When possible, quote only brief phrases and incorporate them into your own sentences.

☑ Set longer quotes off from the text, prefacing them with a brief introduction and following them by an explanation—in your own words—of their main point.

☑ Footnote any words *and ideas* that are not your own—even when taken from the Web. Professors have ways of searching the Internet, and paraphrase is still plagiarism.

Plagiarism Detection Enters the 21st Century

If you're ever tempted to "borrow" phrases or ideas from other authors (especially on the Internet), have a look at the University of Maryland, University College's Center for Intellectual Property Web site. This amazingly comprehensive site will acquaint you with tons of plagiarism detection sources (all available to your professor), as well as articles assessing the prevalence of, and motivations for, plagiarism. Its Web address is: *www.umuc.edu/distance/ odell/cip/links_plagiarism.html*

To see a sample of the sort of service your professor might be using (and how it works), take a look at these sample sites:

turnitin.com/static/resource_files/tii.html (no www.)

www.ithenticate.com/static/tour_master.html

TIP #6: Be Selective

One of the most frequently overlooked factors in producing the perfect paper is the importance of being selective. *Include in your paper only those points that are directly relevant to the question asked or the issue discussed, and eliminate all material that isn't.* As you are constructing your paper, make sure that each point is related—directly related—to what you are arguing at that stage in the paper. Ask yourself, is the point I am making located on the main highway, or is it just some scenic detour? And keep asking yourself as you proceed through the paper.

Students often feel that there are certain things that just *have* to go into their paper—even when they don't see how, or where, these points would really fit in. Maybe the student discovered something really great or unusual in his or her research. Maybe the student thought that the professor had harped so much on some point in lecture that it simply must make its way into the paper. And maybe the student feels that he or she simply doesn't have enough relevant content to fill up the paper (more on this in tip #8, following). Whatever the reason, the result is the same. The student wedges some point into the paper that really has nothing, or very little, to do with the question at hand.

It's never fun to cut out material that you have painstakingly researched or cleverly thought up. But if you feel a sense of discomfort or unease about whether a point is really related to your paper and its goals, this is an important sign that the point really *is* irrelevant. And if as you are writing your paper you feel an irrational impulse to throw in some content (even though you don't really know why you're putting this point in, or even where it should go), then this, too, is an indication that the point should make its way into the recycle bin.

Trust your instincts. What you decide to leave out is often as important as what you choose to put in.

Of course, you might have some information that at first glance seems irrelevant, but which you know really *is* relevant to your ongoing argument. In such a case, you need to explain the relevance of the material when you first introduce it. This will help your reader see the exact role that that material will play in your discussion, and also highlight for the reader the not-so-obvious relation between that point and your thesis or argument. Your paper will seem better argued and more nuanced, thus earning you a better grade.

But what's the big sin in having a little too much content? Why not throw it all in (especially if it's all true) and just let the professor sort it out? Well, for one thing, having lots of irrelevant material detracts from your main argument. It makes your paper lose direction and force because it ties up the reader's mind with lots of extraneous issues. And it gives you less time and space to put in and develop the really good stuff.

Keep in mind that papers are not intended to be comprehensive. While tests might be cumulative, papers usually aren't. In a paper the professor is sizing up your knowledge of a field by asking you to discuss a particular, and relatively narrow, question or issue. The professor is testing by *sampling*, not by covering huge expanses of material. As a result, only certain materials are going to play a role in the paper answer. Part of what's being evaluated is how carefully you select the materials to be included in—and excluded from—your paper.

This selection counts directly in the grade. Some professors (not such nice ones) will simply refuse to consider content that is irrelevant. They simply pretend it's not there. Other (even less nice) professors actually take off for material not directly related to the main argument. So your paper could wind up with huge holes or, worse yet, keep losing points. After clarity, relevance is usually the thing professors care about most. So be selective—and be relevant. Always.

TIP # 7: Move Logically toward Your Goal

A good paper has direction and impetus forward. Like a good symphony, it moves effortlessly through a series of stages, each point naturally flowing into the next. It has an energy, a "pushing forward" that draws the reader along with it. At each point the reader understands where he or she is going to, and where he or she has come from.

When you *outline* a paper—no matter how well your outline captures what the line of argument in your paper will be—your outline has a static quality. Each part of the outline stands as a separate numbered or lettered item on a list. And the outline as a whole is a "freeze-frame" snapshot of the paper as a whole, correct and complete, but devoid of life. As you write, though, you need to create a sense of dynamic motion that propels your argument from start to finish. How to resolve this paradox?

Before you write, look over your outline and ask yourself these questions (and as you write, keep asking them): Why is this point here? What work is it doing? How is this point related to what I'm about to say, and how is it related to what I've just said? Does this point push my argument forward, or is it merely occupying space? Is there an energy throughout the paper that pushes the reader forward? And finally, what is the overall organization of the paper, and how does each point contribute to it?

One technique that is useful in writing the paper is to em-

ploy a good selection of "indicator words"—terms that indicate relations of ideas and changes in the direction of the argument. Words such as "moreover," "furthermore," "in
addition," "next," and "finally" mark *sequence*. Words such as
"therefore," "consequently," "then," and "accordingly" indicate *inferences* or *conclusions* drawn from previous claims.
Words such as "but," "however," "still," "yet," "nevertheless," and "by contrast" establish *opposition* between your
present claim and the claim that precedes. And words like
"now," "indeed," "so too," "of course," and "to be sure" serve
to *resume* a previous point and sometimes to *make it stronger*.

Another useful technique is to take stock from time to
time of what you have already accomplished. Introduce
sentences explicitly summing up what you have shown to
that point: "Thus far we have seen that _____. I turn now
to the question of _____." Students (and some professors)
think such "signposts" are artificial. But in fact they are very
helpful to the professor, who is striving to keep track of what
has (and hasn't) been asserted in a particular student's
paper—often in a stack of 50 papers on the same topic.

Professors want to see sustained—and logically directed—
argumentation in a paper. Make sure you provide it.

WANN'AN (A)?

> The logic of your paper can be a crucial component of
> the grade because what differentiates papers one from
> another often isn't their truth or comprehensiveness.
> In any given class, many of the papers are not going
> to make a lot of mistakes or leave out a lot of impor-
> tant material. One of the things that sets off the A
> from the B papers is how well the points of the paper
> are sewn together to form the case for the paper's
> thesis. Many students have the same data; only a few
> organize it forcefully.

TIP #8: Go to the Limit

One of the first things students notice when they get their paper assignment is the number of pages that the professor is requiring them to write. Normally professors will designate a certain page range, say 3 to 5 or 8 to 10 pages. You might look at the low end of the range and give a sigh of relief. But you can just go ahead and cross that low number out. Strong papers usually go to the limit.

By now you know this doesn't mean that you should fill your paper with irrelevant material, lots of quotes, and all sorts of other B.S. Nor does it mean that you should format your paper with really big margins, gigantic fonts, and— horror of horrors—triple spacing. (Yes, professors do notice such things. They are not fooled and are sometimes even annoyed when they encounter such an obvious ploy.)

On the other hand, we don't recommend that you *exceed* the professor's page limits. Some professors want to see that you can express yourself succinctly enough to cover all the material within the page limitations. Also, extra pages to read means extra work for the professor. Many won't mind some extra pages in a great paper. But in a less than stellar paper—especially one with lots of irrelevant materials that account for its excessive length—the extra pages could lower your grade.

Writing the maximum number of pages helps your paper be a maximally complete and well-developed answer to the question. Professors pay particular attention to such things in their grading, especially when they are separating the excellent papers from those that are only good.

✔+ *EXTRA POINTERS*

If you feel that you don't have enough to say to fill the space, try doing one or more of these things:

- Make sure that you have done all the assigned tasks.
- Explain the points you make more fully and/or more clearly.
- Add additional examples to support your points.
- Probe the central points of the paper more deeply.
- Provide more analysis for your points.
- Examine the relations between points more fully.
- Consider adding another source to a research paper.

Professors' Perspective

You might wonder why professors give ranges for papers anyway. If the A paper pretty much requires the maximum length, why not just tell everyone to write papers that long? Well, professors have to create assignments that are doable by students at a variety of levels of ability. They have to construct an assignment that will address their class's lowest common denominator. In addition, professors recognize that there are some gifted students who think and write so clearly and succinctly that they could actually write a great paper in the minimum number of pages. But most students won't be able to pull this off.

TIP #9: Draw a Conclusion

No paper is complete without a definite conclusion, expressed in a full paragraph—the last paragraph, of course. In the conclusion you should state what you have shown in the paper and how you have shown it. If the paper was conceived in the form of a question, then the conclusion should state a definite answer.

The trick here is not to be too repetitive. If you just go over everything you said, in the same way you already said it, your conclusion will be boring. So sum up only the highest and most original points of your paper. It's especially elegant (though not necessary) to take the paper to a new level in the conclusion. Can you tie together all the issues in a way that brings out something new? Could you place the problem into a broader context by explaining the importance of your result? If so, your conclusion itself could earn you some extra points.

Under no circumstances should your conclusion be used as an occasion just to state your personal opinion—that is, your own beliefs and views, independent of the arguments you've given or the materials you've studied in your paper. Some students think the conclusion is the time to say what they "really think." It's not. Any points that you make in your conclusion need to follow logically from the line of argument pursued in the paper. A conclusion is an end, not a beginning.

★★★★☆ **4-Star Tip**

Some analytical (and some hybrid) papers ask you to critically evaluate an article, or consider two different interpretations of the same issue. It's tempting to end this type of paper in a tie. This is not good. You need to end this sort of paper by taking a firm stance—one that doesn't come out of the blue, but is the bottom line of your whole paper.

Keep in mind also that the conclusion is the last thing the professor will read. It's the moment of truth. After reading your conclusion, the next thing the professor will do is affix a grade to the paper. So use your conclusion to remind the professor of all that you have accomplished in the paper. You want the professor to have that warm feeling of a job well done right before he or she lays down your paper grade.

TIP # 10: Edit—and Rethink and Revise

Writing is an active process that involves writing and rewriting. Rewriting is essential. By rewriting, we're not talking about proofreading to correct spelling and grammar errors. Yeah, you do need to proofread. But the real action here is in going over the paper, really thinking again about what you've said, and being willing to make changes as necessary. Now is an especially good time to do this. Before, you didn't know what you were going to say, and had to spend time figuring it out. Now you know—you've said it.

The professor might also be willing to read your draft, or

to answer an e-mail detailing your main results, before you begin rewriting. If so, you can still take advantage of this opportunity. But even if not, be your own professor. Read the paper as you imagine the professor would, then rethink and rewrite. That's how to advance to the next level.

You Can Do It!

Everyone feels pain in the editing process, even the best writers. It's very hard to rethink and revise your own work. Who wants to do more work when the paper is (apparently) done? Who wants to make changes in the paper you've so painstakingly crafted? Who can even bear to look at the thing without wanting to throw up?

Critically looking over your own paper typically involves some unpleasantness. So here are some techniques that we ourselves find useful in managing the pain of the editing process:

- **Take a break.** Look at your paper in the morning when you have some distance.
- **View the editing process as a genuine learning experience** that will increase your understanding of the material as well as improve your writing skills.
- **Remember how far you've come** and how much you've already accomplished. Don't stop now, you're almost there.
- **Anticipate the pleasure** you'll feel in handing in the paper—and getting a good grade.
- **Promise yourself a small (or large) reward** when you finish this most unpleasant of tasks. (Funny how these rewards always seem to figure in.)

Review Session

Writing a good paper is the culmination of a multistep procedure that takes place over an extended period of time. Understanding the question and the kind of task to be done; doing the analysis and/or research; going to see the professor or TA as many times as is reasonably necessary—all of these are part of getting your ideas in shape to write. But the writing of the paper is, ultimately, a separate activity worthy of careful consideration. For in the end what you're going to be graded on is the *product* you produce—considered on its own—not the *process* you've gone through in manufacturing this product. There's a moment of truth that occurs when the professor picks up your paper, works his or her way through the various things you say, and is pleased (or less pleased) with what he or she finds.

You'll increase your chances for an A encounter if you've followed our 10 tips for constructing the perfect paper. In fact, the 10 tips can be reduced to 5, which will make them easier to remember—and to follow:

1. Get right to the point and advance a thesis.

2. Be clear throughout, and write in simple and direct

language for the reasonably intelligent person—not the professor.

3. Be selective in your points, always moving logically toward your goal.

4. Go to the page limit (though not over) and draw a definite conclusion.

5. Edit, rethink, and revise. Think of your paper from the professor's perspective.

PART 5

THE LAST MONTH

CHAPTER FOURTEEN

The 4 Hazards of the Last Month of the Semester

How ironic. The moment when you think you can finally see the light at the end of the tunnel is the most treacherous time of the semester. Gradewise, we mean. The last month of the semester is rife with hazards. They can throw your GPA into a real tailspin if you're not careful. So now's not the time to start packing the supplies for the end-of-semester celebrations. It's time to hunker down and prepare for those hazards that are coming your way in the waning month of the semester. To master the avoidance maneuvers that can head off those obstacles, and the remedies that will allow you to overcome any difficulties that might occur. So that you can stay firmly on the road to good grades—all the way to the very end!

More often than not, the bulk of the grade in a typical college course is awarded during the last month of the semester. Think about it. You might have to turn in a research paper (worth 25 percent of the course grade), there might be a final exam (25 percent of the grade), and there might even be a presentation in section (worth half of the 20 percent class-participation grade). Add it all up, and the last

Professors' Perspective

You might wonder how professors came up with the brilliant idea of having up to two-thirds of the course grade awarded during the last month of the semester. Sometimes it's because professors don't want the first piece or pieces of work to count too heavily. They know that some students blow off the first test or, not really knowing what to do, mess up the first paper (or two). Professors don't want these students to be too heavily penalized for their first missteps and, more important, do not want to deal with hordes of students who are down on the course after only a month. Finals also weigh heavily in courses where the material is cumulative (think about how much more sophisticated your command of Chinese or Biblical Hebrew or Ancient Greek is in the 15th week of a course than it was in the 4th week). Professors in these sorts of courses do not want the more basic skills learned at the beginning of the course to count as much as the harder ones mastered by the end. And sometimes a course is bottom-heavy simply because the professor is running behind schedule and has postponed the due date for some paper or the date on which some exam is administered. In such an event the professor has unwittingly increased the grade allotment to the last weeks of the semester (and with it, the end-of-semester pressure for his or her unlucky students).

month could count a staggering 60 percent of your grade in the course. This "backloading" of the grade is even more extreme in courses where there's a final exam that counts very heavily. It's not unheard of for finals to count 30 or even 40

percent of the course grade. Some schools even have rules
forbidding professors from counting the final for more than
50 percent of the course grade!

Whatever the reason for the concentration of grades at
the end, the net result for you is the same. It's the last
month of the semester and most of your grade is still up for
grabs. All this, just when those last-month hazards are star-
ing you in the face.

HAZARD #1: Out of Gas
REMEDY: Fill 'Er Up (Your Momentum-Stores, We Mean)

It happens almost every semester, to almost every student.
A few months of the semester have passed. You've taken a
test, maybe even written a paper. It's practically Thanksgiving. Or Spring Break has just come and gone. Things seem
to be going well. Or maybe not so well. And suddenly you
feel like you're out of gas. No energy. No momentum. And
now everything feels like an enormous effort. You dread going to class, still having to write another paper, or take yet
another test—or two. And all you can think about is: when
will this bleepity-bleep course be over?

If you sense that your momentum is practically grinding
to a halt, you need to recharge your battery. To regain that inner strength, purpose, and direction that will be needed to
navigate your way through what's left of the course. It's not
primarily a matter of studying harder, pushing yourself more,
getting better organized, or even making sure you go to class
(though, of course, these might figure in). No, what's at issue
here in most cases is something deeper—and something
more internal. It's about getting yourself motivated—*again*.

You Can Do It!

While we as professors are quite expert at recognizing the signs of an empty tank—we need only to look around at the students during our lectures of the 10th week—we don't have any special training in motivational techniques. Luckily we have friends who do. Like our educational psychology colleague,

Visiting Professor

Myron H. Dembo, University of Southern California

In my book *Motivation and Learning Strategies for College Success* (and in the course I teach on the subject at USC), I show students various strategies for becoming better "self-directed" learners. Here are some of my best ideas:

1. **Set specific goals and monitor daily progress.** The goals you set for yourself influence what you attend to and how hard you try. As the semester moves into its last month, it is important that you focus on those tasks that lead to improved academic performance.

2. **Focus on effort.** When things go wrong, many students tend to attribute their failure to lack of ability rather than to lack of application or of effort. These students are less likely to be optimistic about future tasks. Identify strategies that you can use to improve your academic performance through increased effort—improving the way you study, seeking help from the instructor, and, in general, taking charge of your own learning.

3. **Use positive self-talk.** What we say to ourselves is an important factor in determining our attitudes, feelings, emotions, and behavior. Positive self-talk, e.g. "I can do as well as anyone in this class," can serve as a motivator for trying new tasks and persisting in learning.

4. **Reduce anxiety.** Anxiety can negatively impact academic performance, especially when students are worried they will not obtain their desired grade in a course. Anxiety then keeps students from changing behaviors that stand in the way of their getting better grades. Consider anxiety-reducing strategies such as relaxation training that focuses on controlling one's breathing, as well as meditation, exercising, listening to certain music, and using self-talk to calm you down, e.g. "There is no need to be worried. I can do this."

5. **Combat procrastination.** Many students are procrastinators by nature: they think, why do today what I can put off till tomorrow? This tendency can come to the fore at the end of the semester when projects are larger, and more material is to be mastered. Develop strategies to overcome the difficulty in getting started, and the tendency to view any large task as insurmountable. Some students find the "bits and pieces" strategy helpful: you divide your task into manageable parts, then set as your daily goal the completion of just one of the subtasks. Another strategy is the "5-minute plan." Here you agree (with yourself) to work on a task for 5 minutes. At the end of 5 minutes, you decide whether you will work on it for another 5 minutes. Before you know it, you're working on a task for 30 minutes or longer.

6. **Use external rewards.** Use self-reinforcement to maintain your motivation. This strategy involves making deals with yourself, e.g., "If I outline my English term paper, I'll go out for pizza with my friends," then taking pleasure in those earned rewards.

7. **Change your study environment.** Many students study in inappropriate environments. They simply can't control distractions. It's never too late to leave your dorm room or other familiar locations where you are more easily distracted, and move to a more quiet place such as the library, where you are less likely to be distracted. Research indicates that high-achieving students report greater use of "environmental restructuring" than do low-achieving students.

8. **Join a study group.** Sometimes students get tired studying in the same way throughout the semester and need a change to motivate themselves. Joining a study group can provide the extra motivation to review material by discussing it with other students in the class. In addition, sometimes seeing other people engaged in the same project as yourself can make you feel better about the course and about yourself.

The 4 Hazards 293

9. **Visit your instructor.** If you haven't scheduled an appointment with your instructor, it's good to do so toward the end of the semester. An encouraging or supportive word from an instructor can often push you along through what's left of the semester.

10. **Deal with your emotions.** There is increasing evidence that stress and depression are on the rise among college students. Negative emotions—e.g. sadness, depression, decreased energy, and feelings of hopelessness—are experienced by many students toward the end of the semester. If you experience these negative emotions, seek assistance from the counseling center at your college or university. In addition, negative emotions are often caused by inappropriate beliefs such as "I have to be the best student in the class," or "I'll be able to get an A only if I write something the professor has never thought of." Consider evaluating your beliefs by asking questions like "Where is holding this belief getting me?" or "What evidence do I have to support my belief?" Then get rid of any irrational beliefs you have found.

HAZARD #2: You've Got Too Many Balls in the Air
REMEDY: Learn to Juggle (or Put a Few Down)

In the last month of the semester, it is not uncommon to find you have more work than hours in the day. You're taking five classes and each of them has a final exam and another test or paper left to do. Not to mention that these papers are longer than previous ones, and the final is two or three hours long, not a 50-minute hour like the regular tests. With this much work crammed into such a short period of time, it's practically guaranteed that you'll have two tests scheduled for the same day, or a paper due on the same day as a test in another class—or maybe even three things coming due at the same time. (Professors don't usually conspire to make their due dates coincide; it's just that many divide up the semester in the same pattern, and in addition have no say in when the final is given.)

And if that weren't enough, the upcoming holiday season or summer break looms large—you need to buy airline tickets, get your car in shape, make plans with your friends or family. And soon.

What a nightmare! How are you ever going to juggle all of this? A few simple strategies can make a world of difference:

- **Make a plan**. Get out your calendar and plot out all your deadlines for the last month. Check the syllabi for all your courses to make sure you have complete and accurate calendar entries. Pay particular attention to when the final is given, as it is often at a different time—and in a different place—than the regular lecture.

- **Prioritize (with an eye to the grades)**. Once you've determined what still needs to be done in each course, make your best estimate (or even guess) about how much time will be required for each task. Then map out work sessions (as best you can) to enable you to complete each project on time. Be sure that you allot more time for pieces of work that count *for a greater part of* the course grade (and, of course, less time for pieces that count very little, such as that 10 percent section report). And be sure to factor in the level of difficulty (for you) of each of the pieces of work. Sometimes you need to budget more time for harder items even if they don't count quite as much in the course grade.

- **Don't rob Peter to pay Paul.** Sometimes students who are falling behind in one course simply blow off activities from another course. What's the harm in missing a week of lectures in course A, they reason, if I'm using the time to do the paper for course B? It's obvious that such a strategy accomplishes nothing, because what you gain in one course you lose in the other. But less obvious (to some) is that going to class is one of the *most* efficient strategies when you feel you have too much to do. Not only does the professor offer up the material in a highly processed—and often targeted-to-the-grade—form, it's also not nearly as much effort to simply write down what someone else is saying than to master the material from scratch on your own. Given a choice of a lecture or an hour of work on your own, always choose the lecture.

- **Don't be a perfectionist.** In the last month of the semester, do the best you can, but don't strive for perfection. Perfection is not required for a getting an A on a paper or test. And there are no A++++'s. *Use available clues for determining what the expectations really are on that term paper or final.* Try to think out what depth of analysis is really required on that analytical paper; how much research is really expected for that term paper; and how much material really has to be mastered for the final (and in what detail). And if you don't know, find out. Ask the

professor or TA about the remaining contours of
the course—and how much work might be *too
much*. Look, it's crunch time now. And it's crunch
or be crunched. But . . .

- **Don't cut *too* many corners**. Some students go
overboard taking shortcuts in the last month of the
semester. They try to figure out exactly how many
points they'll need on the test or paper to get the
grade they want. Then they aim to get exactly that
number, but no more. This strategy almost always
backfires. Students aren't usually in a very good position to know exactly how much studying is
enough to rack up some precise number of points.
What happens if you choose to study only 80 percent of the material with the goal of getting an 80
on the test, but the professor focuses the test on the
other 20 percent? And in any case, there isn't that
direct and exact a correlation between mastering
the material and getting points on the test. Sometimes the final grade depends on just what material
is chosen to be tested; how well the student constructs the essay (not just how well he or she has
prepared); or simply how well (or badly) the exam
goes. Don't cut too much, and don't cut it too finely.

- **Shed commitments**. This is the time to avoid any
extras. If you are doing optional work for your
courses—such as recommended readings (or sometimes even tangential, *required* readings)—it's time
to let that go. And if you have demanding hours at
a job you hold, you might seriously consider cutting back your work hours—if at all possible.

- **Avoid distractions**. Maybe this is not the best time to: pack up all the worldly possessions of your boyfriend or girlfriend and toss them out of the second-story window; wean your five-year-old kid; train your dog to jump through hoops of fire; engage in elaborate roommate swapping; or decide to start training for the Boston Marathon. Save these activities for after the semester.

★★★★☆ **4-Star Tip**

There are many *academic* activities that arise in the last month of the semester that, even though they have value, are also great distractions and can be very time-consuming. Avoid such bottomless pits as: investing large amounts of time in preregistration, researching or picking your major, planning to study abroad, looking into a summer internship, or planning to transfer to another college. These are all wonderful things to do—just not now.

HAZARD #3: You're Just Not Getting the Grade You Want
REMEDY: Take It to the Next Level

Sometime in the last month of the semester you might experience the dawning (and possibly upsetting) realization that you're not doing as well as you'd like in some (or all) of your courses. Sometimes there's a course that seemed to be going great, but still you're only getting a B in it. You keep thinking the *next* assignment will net you an A, but now you realize that there aren't so many "next assignments" left—and that if you don't crack an A soon, your chances for an A in the course are going to go down the tubes. Or maybe there's a course in which you thought you were doing okay, but now that you think of it, you're really on the track for a C, not a B. In which case some emergency repairs are in order.

If any of this is happening to you, it's time to take stock. To size up the situation. See where you stand. Right here and now (while there's still plenty of time).

Gather together all your graded work—you know, the stuff with your professor's scribbles on them. Not just the tests, but the quizzes and papers, and the graded homeworks, problem sets, and lab reports, too. Make a directed search through each piece of work for the answers to these questions:

- Where did you lose points?
- Was there any particular *kind* of question you had difficulty with, or did you just lose a few points here and a few points there?
- What sorts of negative comments did the professor write in the marginal notations or the summary comments of your work?
- Do the same kinds of criticisms come up on each homework, quiz, test, or paper, or are there completely different comments on each?
- What would you say is the single *biggest* reason, as reflected in your previous graded work, that you are not getting the grade you want?

Mark up your work. Take notes if you have to. Because how successfully you've answered our questions above will be an important determinant of how well you will do in the rest of the course.

But it's not enough to simply survey the causes of your less-than-hoped-for performance. What's needed, in addition, is a diagnosis of (and a repair plan for) the *causes* of your previous difficulties. Did the problems arise because you didn't do the reading or cut class? In that case, resolve to make it to all the remaining classes and to do at least the central reading. Was it because you couldn't understand the material or didn't know what the professor wanted? Do all the activities of the course in the order they were intended (reading, lecture, and section), and use all the clues (by now there should be many) for what the professor is going to be testing you on. Did your actual work during one or another exam not go far enough, or did you simply get flustered or panic? Perhaps better preparation is needed, or maybe you need to budget your time better or construct a fuller or clearer answer to the exam question.

Whatever you determine the cause of your problem to be, make a plan for corrective action. *Self-diagnosis will have no value if it's not followed by self-change.* Ask yourself what specific changes you can make to remedy the specific problems you've located in your previous work. And answer yourself honestly. No point lying to yourself.

You Can Do It!

Sure, we know that this kind of "reality check" might not be fun. No one likes to look over past inadequacies—or even just momentary lapses or errors. But if you want to change your grade, you have to be fully clear about where you are now. Every diet book has you get on the scales before you start. Every exercise book has you assess your physical condition before you lift any weights. The good news is that if you do this now, at the beginning of the last month of the semester, there's still plenty of time to turn things around. Here's a time when the backloading of the grades can really be turned to your advantage.

Keep in mind that often it doesn't take nearly as much as you think to make really big improvements in your grade. Provided you make *targeted* changes—ones based on the specifics of your previous performance in the class and on what is being counted for the grade—a little effort can have a huge payoff. And don't view this as a time to improve every study skill you have. Usually, you need to make only one or two strategic changes to move your grade up to the next level.

Need some more help figuring this out? Here are our tried-and-true guidelines for moving from a C to a B, and a B to an A.

How to Move from a C to a B

If you are getting a C in a course, this means that the work that you are producing includes *real mistakes* or *significant omissions*—or, in the bad case, both. The first step in getting out of the C range is to make sure you have the basic information of the course correct and complete. And to get to the point where you clearly understand these materials—so clearly that you can communicate the materials in a way that someone else can understand what you are talking, or more likely writing, about. But you don't need to have a very sophisticated grasp of the course to get a B. You don't need to present elegant and embellished renditions of the course content. You need just to get it right and to let your grader know (without the slightest doubt) that you *have* gotten it right. When you can do this, then you'll get your B.

To get to the B level, it helps to locate and distill the few main points of the course and nail down a solid and complete understanding of them. Often C students are the ones who get bogged down trying to learn every little piece of information. So bogged down that they never learn the basics. Professors regularly see C students in the last month of the semester who come in asking about all sorts of picayune issues. After a few minutes of discussion the professor realizes that what's really causing the difficulty is that the student hasn't understood some basic concept that was covered in the first weeks of the class. The professor explains that point and all of a sudden the course finally falls into place for the student. The student's face lights up with the realization that he or she finally gets it—and that it wasn't nearly as complicated as he or she had thought.

So, if you want to get from a C to a B, start by going back to the basics. Look back at the syllabus to see the overall organization of the course and the course goals. Look over your lecture notes to see what topics the professor has been covering in the course. Look over any handouts that might have come your way in the lecture or discussion section. Then try to identify the few (two or three) points that are the main focus of the course. And what other points flow from these. And be sure to see how these main points were tested in the various exams and papers to date. Once you know what the key points are, you'll have a much easier time detecting how the professor ensured that you had mastered these very points.

This also wouldn't be a half-bad time to consider getting some extra help from the professor (or TA). Office hours are often especially undersubscribed in the weeks after the midterm, and the professor, not having to worry about an immediately upcoming paper or test, will be happy to talk to you at some length about whatever is giving you trouble.

In addition, many students, especially in problem-solving courses such as statistics, logic, physics, or chemistry, can benefit from work with a tutor (either paid, or free at the learning center of the university). These tutors can spend more time helping you catch up than could the professor or TA. You might be more motivated to do the requisite work if you're directly accountable to some individual—perhaps even paying him or her. And sometimes some encouragement and handholding from someone not affiliated with the course can bring the confidence that results in great grade benefits.

After going through all this effort, don't forget to apply what you've learned to all the remaining assignments in the course. Keep in mind that you've got to break away from what you've been doing in the past if you are going to make improvements. So actively fight any inclination to go back to your old and comfortable, but in the end not successful, ways of looking at the material and of doing things.

Once you've moved from a C to a B—or if you're already getting a B—you might consider moving from a B to an A. Hey, things are going great. Why stop now?

How to Move from a B to an A

If you are getting a B in a course, you're not doing anything particularly wrong. You have the information essentially correct and complete, and have shown the professor a good mastery of the material. But there's nothing at all *distinguished* about your work. Your papers and tests read like a dozen other ones in the stack. The professor reads them and thinks they're fine, but they don't excite him or her. They don't have any special "lift." They don't stand head and shoulders above the pack.

To get an A you simply need to make your work better. On tests you can't give up as many points. Some students get B's on tests by being just a bit careless. They lose 2 points here and 2 points there. Maybe they didn't try to memorize everything they needed to, or missed a few classes, or didn't bother to check over their answers. Unfortunately sometimes just a few points, maybe as little as half a point, can make the difference between a B and an A. So tighten things up. Don't give yourself the luxury of dropping any points that a little extra thought or care would capture.

Getting B's on papers (or sometimes on essay questions on exams) usually means that you didn't probe hard enough, push far enough, or investigate in enough depth some key issue in the paper. Perhaps there's a particular part of the analysis that stopped short, or a particular source whose significance wasn't considered thoroughly enough. The paper runs out, the professor thinks, just as it's getting interesting and pushing into really excellent territory. And—if they are doing a decent job writing comments—professors usually will make some marginal notation (often repeated in the summary comments at the end of the paper) about the exact location at which more thinking, more analysis, more probing could have been done. So go over your paper comments and try to pinpoint any junctures where, if you had pushed further, you could have broken through the barrier between the B and the A grade. Look for the exact point your paper could have soared.

In other cases what distinguishes the B from the A paper is that B papers are too "flat." The thesis is too obvious, the examples too simple, the relations among the various points aren't explored with enough insight. For a professor, these sorts of papers are just ho-hum (some less nice professors might describe such papers as boring, ordinary, and surely not A-level work). Getting these papers up to the next level requires more imaginative and more creative ways of processing the materials. Not just *more* stuff, but a better or more original way of looking at *the same* stuff. If this is what seems to have been the problem in your paper, then it would be especially helpful to meet with the professor (or TA) and talk about what else you could have done in the paper, and how you could have explored the topic or answered the question in a more creative way.

Again, be sure to take the insights you gain from assessing your work and from talking to the professor, and make adjustments in your next papers (or tests). And stay positive! The fact that you are coming so close shows that you are almost there.

HAZARD #4: You Stand a Real Chance of Failing
(If You're Not Careful)
REMEDY: Fish or Cut Bait

But sometimes the situation isn't so rosy. Here it's not just a matter of moving from a fair to a good grade, or a good to an excellent grade. It's a question of whether to stay in the game at all. This occurs for students who find themselves, toward the end of the semester, looking at a solid D (or worse) in a class.

If you find yourself in this position, you have a serious choice to make. You might consider dropping the course. The drop deadline varies from college to college, but at many schools it extends into the last month. At the University of Arkansas, for example, you can drop without grade penalty up to the 12th week. (Check with your adviser for the rules at your school.) On the other hand, you can decide to stick the course out. You'll have to do the best you can, and aim for a decent, if undistinguished, grade. Probably the best you can hope for is a C, though maybe a B is still possible. Before you decide what you're going to do, consider both sides of the question:

Is It a Good Idea to Drop a Course?

PROS

1. You'll get out of something that's not all that likely to end well.

2. You'll get out of something that is upsetting you and might continue to upset you.

3. You will free up 20 or 25 percent of your time to devote to getting good grades in your other courses.

4. You will protect your GPA (dropped courses don't usually count in the GPA) and maybe even ultimately improve it.

5. You could get a good feeling from taking a positive action to remedy your situation, or a sense of relief from getting out of the killer course—which might help you do better in your other courses.

CONS

1. If the course is required, you'll have to take it again and there's no guarantee it'll go any better next time.

2. If the course is a prerequisite for another course, you'll have to delay taking that other course (and all the other courses that depend on it) till you've retaken the first course.

3. Dropping courses late in the semester is a bad habit to get into, and after you've bailed out once, it can become easier to do it again.

4. You might end up short on credits for advancement to the next academic year, or for graduation.

5. You might feel a sense of regret or depression about giving up. Especially after you hear how easy the final was.

If you're seriously thinking of dropping a course, it's always a good idea to inquire of your professor or adviser before doing the deed. Sometimes he or she can give more information about where you stand, or what's to come in the course, that might centrally affect your decision. And of course, it's often useful simply to have someone to "bounce your ideas off" before you make such an important decision about your program.

If you do decide to drop the course, be sure you've made your decision in the cold light of day rather than in a moment of passion ("I've got to get out of this thing, no matter what the cost"). And make sure you have given some thought about how you're going to make up the lost credits and satisfy the requirement (if any) that the course was supposed to fulfill. Also make sure you withdraw from the course in such a way (if possible) as to get a "W" (a late drop that doesn't figure into your GPA), rather than an F (which, of course, will enter into it).

But if you decide to stick it out—brave soul that you are, or just past the drop deadline—buck up! We have tips for you, too, which, while not *guaranteed* to work (hey, we're professors, not magicians), at least will give you a fighting chance for salvaging your grade. Here is our 10-step plan for radical grade repair:

How to Move from a D to a C (or Maybe Even a B)

1. **Don't kid yourself.** You're doing badly, really badly in this course. So you need to undertake a plan for radical grade repair. (We've found that simply admitting the problem is often the hardest step for students in difficulty.)

2. **Make a plan with the professor.** This isn't the time for do-it-yourself repairs. Make a real appointment with the professor (or TA) in the course. Explain to him or her, seriously and in some detail, what problems you're having (bring along your graded work to date). And ask what specific suggestions he or she would have for the remainder of the course.

3. **Work from where you are.** Students who are having serious difficulty generally do best if they concentrate their efforts on the material from this point on, rather than trying to start again from the beginning of the course. Look, you're not aiming for the perfect score, you're trying to salvage as best you can what's left. (In some cumulative courses, this approach will not work and you might be forced to drop the course if you are far enough behind.)

4. **Renew (or in some cases create) your commitment to doing all the course activities.** Resolve to conscientiously do all the remaining course activities—and in the prescribed order. Each week do the reading before lecture; attend lecture and spend the entire time taking good notes; go to the section (if any) prepared; and track your progress at the end of each of the remaining weeks. (Have a look back on our Chapters 5 and 6 for full details.)

5. **Make use of as many supplemental resources as possible.** Consider joining a study group, finding a study buddy, or going to the college learning or writing center. You might even consider hiring a private tutor (graduate student, former TA, or even "whiz kid" undergraduate). If you get someone who really knows his or her stuff (and knows how to communicate it), you can turn things around in a hurry.

6. **Make up any missed work.** Very often the reason you're getting a D (or worse) is that you've failed to hand in some graded piece of work. (In the worst case, you've gotten a 0 for not handing the piece in, where you would have gotten a 59 had you handed anything in at all.) Go see the professor (or TA) and ask as politely and apologetically as you can whether there is any way you could make up the piece you missed—even at this late date and even with a grade penalty. Hey, even a few points are better than none.

7. **Broach the possibility of "extra credit" or "do-overs."** In some courses, professors are willing to offer grade-repair possibilities, such as rewrites of papers, retakes of tests, or extra-credit projects of various sorts. It can be worthwhile to ask the professor (privately and in his or her office) whether he or she offers any of these options. Be sure to raise the subject in an oblique and open-ended fashion. Ask, "Do you ever allow rewriting a paper?" or "Are there any extra projects I could do to improve my standing?" Some professors bristle at such requests, seeing them as seeking favoritism or as otherwise inappropriate in a college course.

8. **Study harder.** Throughout this book, we've emphasized that getting good grades isn't usually just a

matter of studying harder. And it isn't—even for you. Nonetheless, when you're getting a D (or an F), sometimes studying harder—you know, more hours and more intensely—can work wonders. 'Nuff said.

9. **Follow (to a T) our suggestions for preparing for, and taking, the final.** Read on to Chapter 15, in which we offer 17 strategies for acing that all-important final exam. (And if a paper is still left, go back to our method for writing a bang-up paper, in Chapters 10 through 13.)

10. **Maintain a positive attitude throughout.** If you are really to do better, you're going to have to create, and maintain, a positive, "can-do" attitude throughout the balance of the semester. Instead of telling yourself, "I've really messed this thing up," "There's no way I'm going to pass this course," or "Things never go right for me," say to yourself—often and with conviction—"I'm going to get at least a C in this course," or "I've sought help, and I now know what to do," or "Things are different now—I'm really learning well, and this change is going to be reflected in my grade." Defeating your defeatist attitudes is an important part of defeating your defeat!

Review Session

For as long as there are classes—and as long as there are students in these classes—there will be hazards in the last month. Not because anyone is at fault, but because that's just the nature of the beast. Semesters are long—12, 13, sometimes even 15 or 16 weeks. Classes can be boring. People's motivations wane. Things don't always turn out as planned. And grades can be a genuine source of worry—and frustration—for many students. Nonetheless, there are four predictable hazards that come up for many students in the last month of the semester, and 4 specific remedies that can be employed to avoid these hazards (or at least deal with them once they occur):

1. **Out of gas?** Set goals, focus on effort. Use positive "self-talk," reduce anxiety and/or procrastination, and offer yourself rewards. Change your study environment, join a study group, or seek support from your instructor. And deal with your emotions, especially if they're irrational or inappropriate.

2. **Too many balls in the air?** Make a plan and prioritize with an eye to grades. Never cut one class just to work on another. Don't be a perfectionist (but at the

same time don't cut too many corners). Shed commitments and avoid distractions (even academic ones like picking a major or planning a junior year abroad).

3. **Not getting the grades you want?** Take stock and diagnose the problem. Figure out how to produce the basic, correct and complete answers, if you're getting a C. Devise strategies for creating more probing and more creative answers, if you're getting a B.

4. **In danger of failing?** Take radical steps to improve your grade: Admit the problem and make a plan. Work from where you are by doing all the class activities and making use of any supplemental help available. Make up missing work and raise the possibility of extra credit. Study harder, prepare well for the final, and keep your head up throughout.

Sure, there are hazards. But they needn't slow you down. *You* know what to do.

CHAPTER FIFTEEN

17 Strategies for Acing the Final

It's 10 minutes to midnight. At long last the final exam period has arrived. But there's no need to go off the deep end. Because what you might have thought was a total nightmare is really your biggest and brightest chance to shine. At this point in time the stars are perfectly aligned. You know what to do and how to do it. Your professor is eager for you to succeed—now more than at any other time in the semester. If nothing else, he or she wants that final grade roster to come out looking pretty (and will dispense extra doses of help to make it happen). And, in a true embarrassment of riches, we're brimming over with killer strategies for acing the final exam. So relax. Take your shoes off. Look over our tips. And start making plans to get an A on your final—and with it, your course!

STRATEGY #1: Start Studying the Last Week of the Semester (If Possible)

Often there's something of a lull in the last week of the semester. The reading for the class tapers off (or might disappear altogether, if the professor is planning to spend class time catching up); the section meeting has nothing new to say (but is set aside for a general review or practicing sample questions for the final); and sometimes, if you really hit it right, the class is canceled altogether (because it's snowing or the professor doesn't want to start a new topic). And you, reasonably enough, think it's a good time to take it easy, have a break, and get rested for the final.

Bad ideas. All. A much better idea is to get cracking on the final preparation the week *before* exam week. You see, if you confine your preparing to the actual exam week (as many students do), you'll inevitably find yourself having to take an exam and prepare for another exam—both on the same day. Here's how it might go. You have a chemistry exam from 7:30 to 9:30 A.M. (or even till 10:30, depending how long exams are in your school). Get back to your place at 11:15, time for breakfast/lunch. Relax for an hour or two. And now it's 2:00 P.M. You're dead tired—and somewhat depressed, too. You keep going over those chemistry problems again and again in your mind (more on this later). And then—it's time to start preparing for your Russian History final!

Better idea? Start preparing for your exams the week before exams. When you don't have to write exams and prepare for new exams on the same day.

Another advantage of doubling your final exam week is that this will give you a leg up on the competition. A recent study shows that most American college students spend between 10 and 14 hours preparing for *all* their exams. Divide it out, you get two or three hours for each exam. Not too much, given the 40 hours of lectures and the who-knows-how-many hundreds (or in some courses, thousands) of pages of reading you're supposed to master. If you can space out your studying over two weeks, you'll simply have more hours to prepare for each of those precious finals. And everyone knows that anyone *could*, if they wanted, study for more than a couple of hours per exam.

Also, the time you put in studying during the last week of the course will pay off more than the same amount of time invested elsewhere in the course. What you study now will stay in your memory much more than what you studied in the third week. You know now much better what *to* study for the final. And if you've made any progress in the course, your skills have been honed to a point where you can study better than at any previous time in the course.

Starting to prepare a week in advance brings with it another, hidden benefit as well. At just the time when your cohorts are *starting* to prepare—and freaking out—you're two-thirds of the way done and feeling upbeat about the final. A confidence that carries over to the actual exam and enables you to write a well-thought-out and carefully constructed exam.

So when it comes to exam week, make yours a double. Start studying as early as you can. Make the last week of classes *your* finals week.

STRATEGY #2: Blow Off the "Float Trip"

You wouldn't believe it, but in some schools the powers that be (either in the school administration or in various student organizations) plan special activities during the last week of classes, or during the study days before finals. We've heard that some schools set up "float trips"—pilgrimages down the local rivers on inner tubes, canoes, or floats—designed to help students relax before exams (at least before spring exams). Around finals you might see posters plastered around campus announcing these events. And since these are "official" trips, set up by reputable college organizations, you might think these would be good things to do. Part of a nice college experience.

Our advice: fugettaboutit. Sure, these would be nice experiences. And it's great that your college offers them. But you'll have plenty of time for "float trips" and other recreational activities *after* your finals. Spend your time studying, not floating. And not to worry, you'll have plenty of rapids to run during finals week itself—without ever having to leave campus!

STRATEGY #3: Go to Review Sessions (but Only *after* You've Started Studying)

Around finals your professor might very well announce that he or she will be holding a special *review session* for the final. Review sessions are designed to help students prepare for finals (and also provide an easy way for the professor to manage the hordes of students needing help around this time). *You should make plans to attend any review session or sessions.* And not just for the purpose of getting a signature for your fraternity, sorority, or sports team to prove that you went. No, you need to go because now is when the professor is going to offer up the good stuff.

Professors often come into the review session well prepared. Sometimes they will give a detailed presentation going over the main course material. Since review sessions usually don't have an exact ending point—and can last an hour or two, sometimes even longer—this part of the review session could be the best, condensed version of the course material that you're ever going to see. What's more, at the review session the professor often takes the opportunity to make explicit suggestions about what will actually be on the final, or what would be particularly good to study for the final. And even when he or she doesn't, you can often glean useful information simply by seeing what he or she emphasizes (and as important, what he or she downplays or even omits altogether) in the presentation.

Review sessions also typically include a period in which students get to ask any questions they might have about the course (or sometimes even about a particular study question, if such questions have been handed out). That's why it's a very good idea to start to prepare *in advance* of the review session. You can come in with specific questions about things you don't understand, or want to know about, and the professor will be willing to answer your particular questions no matter how narrow or focused. Also, preparing for the review session is a good way to trick yourself into starting to study earlier. If you feel pressured to prepare a question or two for the Thursday night review session, you'll also feel motivated to start organizing your thoughts the week before the actual final.

STRATEGY #4: Go to Extra Office Hours

Some professors hold extra office hours instead of, or in addition to, review sessions. Office hours can also be useful to attend before the final, as long as you go *after* you've started to study for the exam and figured out what areas you're having difficulty with or are not sure about. Same reason as before: tailor-made help. And going to office hours can not only motivate you to study, it can humanize the whole finals experience by providing a fellow human being (the professor or TA) to ease you through this difficult period. (On the other hand, if when you get to the office you find more students than a herd of wildebeests during migration season—or a group session in which the questions being asked by your fellow students are no better than those that would be asked by the wildebeests—you might decide that your time would be better spent studying in the privacy of your room or the library. Which it would.)

✓+ *EXTRA POINTER*

If you're going to an extra office hour and if your course has study questions, consider writing out your answers to those questions and bringing them along to the meeting. At this point in the course—when questions tend to be longer and more complicated— some professors are willing to look over your actual answers to the questions and make suggestions about changes and improvements that can be made. This might not always work. But then again it might.

STRATEGY #5: Don't Spend Too Much Time on Courses You've Got "Locked Up"

It's natural to want to study extra hard for the courses in which you are doing well. The ones that interest you. The ones where you already have an A in the bag. If you feel this, you need to fight this tendency and put a significant amount of time into courses in which you are struggling or having difficulty. After all, when it comes to your GPA, each course counts more or less equally—hard ones and easy ones alike. (Yeah, we know that your Arabic or Chinese course counts for 5 credits. But you get the point.)

We know that it would be more fun to spend the week engaged in the math problems you can't get enough of, or the French literature that gives meaning and purpose to your life, or the journalism class that's preparing you to be a network anchor. But let's face facts. You have only a certain amount of time to devote to studying for finals. So if you can steal an hour or two from the courses in which you're pretty much guaranteed an A, and invest that extra time in some of your tougher courses, that's a significant reallocation of time that will pay off when your grades for the semester get totaled up. And don't forget to pay attention to what percentage each final counts in the total grade for that course. That, too, should have an effect on your allocation of study time during finals period.

STRATEGY #6: Make Full Use of "Advance Information"

By the end of the semester you have a tremendous amount of information available to you about each of your courses. Just think about all you've got now. You've got all the tests your professor has already given in this class, so you should have a pretty good read on your professor's test-construction style. You have all the comments that the professor has written about your individual work. So you should have a decent idea of how your professor is responding to your work. You may also have a study guide to the final exam. Or perhaps you've gotten (legally) an old copy of that professor's previous final in this course. Or maybe you've found someone who took this course before with this instructor and can tell you exactly what this professor always asks on the final. All of this is good to know. And to make full use of when studying for the exam.

But expect to get even more information on top of this. In the form of "dropped hints" by the professor. Immediately before finals—in the last classes, the review session, or the finals office hours—is the time when the professor is most prone to having loose lips. When he or she is most willing to give more information, more tips, more help. The

Professors'
Perspective

You might wonder why professors drop so many hints around the time of the final. Well, one reason is that professors sometimes come into finals season knowing that the grades in the course aren't all that hot. Yikes, they think, I just can't hand out this many bad grades. I'll have masses of students coming to complain about their grades the minute they open up their e-mail from the registrar—just the moment I'm looking to put this whole course to bed. Also, dropped hints are a pretty attractive strategy for professors who feel that there are some parts of the course that they didn't teach all that well. Or some parts of the course that ended up being too rushed. For these professors, a few well-planted hints could help assuage their guilt and also make up for any problems with their teaching. But mostly it's just a matter of pity. Professors look around, especially in large lecture courses and in big review sessions, and see dozens of students lost in the fun house. What harm would it be, the professor figures, to give them a glimpse (or two or three) at what I just this morning put on the final?

professor knows that it's time for last licks. That the closing bell is about to ring. That a big chunk of the grade is still left for the picking. And the professor wants to seize these last moments, these last chances to impart learning, and also to help generate good grades. Be on red alert for these hints: you will not believe how many there are, and how much they can help you in acing that final.

STRATEGY #7: Manage Your Stress—Your Way

As a college student, you have a high-stress job. And the final exam period is the time of maximum stress. After all, the final is the item that (in most courses) counts most toward your final grade. You have to take four or five finals in a very short span of time. Plus you have no control over the times of the exams, and might get stuck in some weird time, like 7:30 A.M. or 4:00 P.M., that goes totally against your biorhythms. And if that weren't enough, it's the time of the year when you naturally would be thinking about other sorts of things, like that soon-to-be-arriving break.

Keep in mind that this is *your* high-stress season. Simply recognizing this, is the first step in being able to handle the stress.

Different people have different ways of managing stress. By now in life you should have an idea of what things help *you* relax. Maybe you meditate, pray, talk to friends or family, go for a walk, or exercise at the gym. Whatever it is, don't let it fall by the wayside. Do some activities each day that are relaxing for you. Don't think you are too busy to take time off for relaxation. Even a few minutes of stress-management time will pay huge benefits for your state of mind—and your exam performance. With one exception—

STRATEGY #8: Take It Easy on the Substances

Many people at times of stress turn to various substances to help them get by. Could be caffeine or energy drinks, alcohol, tobacco, sugar, or other mood-regulating substances (both prescription and free-market). College students are no exception. If it's your practice to use any or all of these items, try to take it easy. Use them only in moderation. Remember that you have a week or 10 days of concerted work ahead of you. It's difficult to construct a good final if you're hungover, zoned out, or climbing out of your skin. And if you're not careful, you'll wind up in one of these states by the end (or even worse, the middle) of finals week.

STRATEGY #9: Treat Your Machine Decently

No, not your car or your iPod. Your body. Remember, you're not just taking your mind to the exam, you're taking your body with it. And, like any machine, your body functions well only when provided with basic necessities. Things like sleep and food. Sounds simple enough, but some college students think there's an advantage to forgoing sleep during the crunch of finals; or they are too rushed and stressed to eat well, or at all. But you can't produce a good product if you haven't slept and haven't eaten. Don't forget that the professor isn't going to grade you on having survived an all-nighter (or two or three or six), but on having produced an excellent exam. *Which depends as much on the state you're in when you write the exam as on the amount of preparation you've done for that exam.*

It's also important to try to stay in good health through the exam period. Naturally, this is the time when students decide to start passing every virus known to man from one to another. It's hard to produce a good exam when you're sick, so hoof off to the student health service at the first signs of illness.

STRATEGY #10: Shed Commitments

Finals last about a week. And that's all. So surely you can manage to jettison some commitments or delay some obligations for one week. If you're working, try to trim back your job. Maybe your boss will give you some extra time off and you can make it up around Christmas or New Year's. Or if you're a parent, get a baby-sitter or relative to relieve you of some of your child care obligations (don't worry, your mother-in-law won't kill the kids). If you have obligations at your church, synagogue, mosque, or temple, maybe someone else could handle them this week (you could do makeup work for the Lord next week).

Of course, you can sit back and tell yourself that you're stuck with all these things besides finals. Or you can get creative. And ask for help. Don't think that somehow you're "just a student" and your needs aren't as important as those of someone with a "real job." Finals are a key time for you. If you explain their importance to your friends, relatives, and bosses, they shouldn't have trouble understanding. And helping make it possible for you to clear some extra free time.

STRATEGY #11: View the Final as a Work Session (Only This Time One You're Fully Prepared For)

As you know from Chapter 8, exams go best when you redefine the experience. When you stop seeing them as problems to be confronted and instead see them as opportunities for you to shine. Chances to show the professors (and yourself as well) how much you've learned and how much you've really accomplished. This holds particularly true for the final.

Think about it. You can do especially good work at the final because by this point in the course you have a much better sense of the kinds of questions you'll be facing. A much better sense of the plot of the course. And now you have much better abilities in the course skills. You're more familiar with the methods of the field and its methods of analysis. You're no longer a beginner. So you have more confidence and can present your answers in a more confident manner.

Of course, be sure to bring a beverage (or two) to the exam. Everyone works better when they are hydrated and awake.

STRATEGY #12: Make a Plan—and Stick to It

It's longer this time. Your previous tests in the class were usually about an hour. But finals can stretch out to a full two or three hours. That's why it's much more important to use good time-management strategies as you work your way through the final exam.

Spend a few extra minutes picking the questions you're going to answer—if you're given choices on your final. In your previous short tests it was important to pick quickly because you might have had only 15 or 20 minutes for the essay. But on a final you could be making a choice about a topic that will consume a full hour (or more) of exam time. So take the time to pick wisely. To weigh the pros and cons of the various choices before making a final decision. And once you've chosen, spend a few more minutes jotting down an outline. You have enough breathing room in the final to allow for some extra prep time.

At the outset of the exam you should also sketch out how much time you're going to spend on each of the main sections of the final. Your plan needn't be written in stone or frozen in thought. Just a working idea. *Make sure that you divide your time proportionately to the point values.* Especially if there are lots of different types of questions to answer, each counting for a wildly different number of points. You don't want to obsess on some fact you don't remember, which counts .7 points—thinking, wait, wait, I'll remember. Even if you finally *do* remember—which you will—it's not worth wasting time on something that counts so little in the big scheme of things.

✓⁺ *EXTRA POINTER*

> Keep in mind that, on many exams, the questions are independent of one another and you need not do them in order. If this is the case on your exam, consider starting with the parts that are easier for you. That way you'll build up your confidence as you go (who wouldn't feel better after killing the first question?). You'll build up a "time reserve" (you work faster on what you know, so you'll have more time left for the other, more challenging questions). And you won't fall into that sinkhole of a question that you're determined to get right, no matter how long it takes (and how much time you take away from the other questions).

All of this requires that you *plan*—and *track*—your time. No matter what, you need to stop work on a question at the time when you need to move on to the next one. This isn't the SAT or ACT, so no one's going to insist that you put down your pencil at the end of each section. No, you're the tester and the testee alike. Some professors will gently help you stay on pace. Jeremy draws a lovely five-foot-high clock on the board and moves the hands as the time progresses.

Lynn announces at regular intervals that it's time to be moving on to the next question. But plenty of professors leave it to up to you to pace yourself. They leave the exam room while you sweat it out, and return just in time to tell you that there are 10 minutes left. Staying on track time-wise will keep you from hearing the 10-minute warning when you still have two pages of an eight-page exam left to complete.

STRATEGY #13: Construct Your Essay Answers Well

Many times an essay question on a final can be more like a paper than a test. That's because the longer time frame allows the professor to ask for a longer essay—one that might be more comprehensive, draw on more resources (readings or analytical skills), or have more parts and subparts than an essay on a midterm or regular test. Some professors will have given out study questions, or review sheets, in advance of the final. In such a case they will expect well-argued, well-presented, paperlike answers (not off-the-cuff, reactive things you've thought up only at the exam itself).

So final exam time could require you to dust off those *paper-writing* skills (have a glance back at our tips on papers in Chapters 10–13). Here are our leading suggestions:

- **Fill the space and the time**. (Fuller and more developed answers usually get better grades.)
- **Be relevant and answer the question head-on**. (In our experience, more students lose points on essay exams for not answering the question asked than for answering the question incorrectly.)
- **Have a thesis**, that is, some point you're arguing for and which the body of your essay is devoted to

proving. (Professors always notice essays that have no thesis and will never give an A to an essay without a thesis.)

- **Make sure your essay has a structure**—that it is not just a laundry list of points—and that your claims follow one another in a logical order. (This is one of the main differences between B and A exams.)

- **Be specific, and offer detailed analysis or examples** of your points. (Professors are a lot less tolerant of generalities in the longer format of the final.)

- **Be sure your essay is really clear**—to the level that someone who hadn't taken the course could understand, just from what you say, what your answer is. (Don't rely on your professor to fill out your explanation, or to understand the answer better than you express it.)

- **Come to a definite and strong conclusion**. (Give your professor that warm feeling of a job well done right before he or she decides your fate.)

STRATEGY #14: Make It Easy for the Grader to Give You an A

As you construct your answer, keep in mind that the grader is going to be very rushed when reading your exam. More rushed than you can imagine. During the regular semester the professor takes a week or two to grade a set of papers or exams. But for finals, at many schools professors are bound by the "48 hour" rule. This means that they have 48 hours after the administration of the exam in which to submit the grades for the course. That's right, two days in which to grade the final exams *and* calculate the final course grade. And that's for the two (or three or even four) courses that the professor is teaching that semester.

What does this mean to you? It means that in many cases the professor has to "glide" his or her way through your exam, rather than read it super carefully. Because of the short deadline, professors normally don't write any comments on exams. They simply can't afford to stop for comments or for prolonged decisions about grading. Reading and grading the exam needs to be accomplished in a few short minutes.

Knowing this, you can see why part of constructing your answer should include showing clearly where each answer starts and where it ends. And why you should carefully number your answers and subparts of answers (corresponding to the numbers on the test) and do the questions (and any subparts) in the *order* given. It's very easy for graders in a rush to get confused about which question the student is answering or to miss parts of answers.

If yours is an essay exam, you should be sure that the first sentence of your essay contains the answer to the question. Not all the details, just the answer—plainly and simply expressed. You don't want to slow your professor down, do you? Once in a while, if your answer starts really strongly (especially if it's just a one-part question and the professor is somewhat pressed for time), he or she will just slap an A on the exam and flip to the next one in the stack.

If yours is a problem-oriented exam (as you might find in a math, physics, chemistry, or logic course), you should be sure to write neatly and show all your work. Many points have been lost when, in the haste of grading, a professor has simply misread a number or variable and taken off points for the whole rest of the problem or proof. And in many cases, a professor will give you points—lots of points—if the work shown is a good start on the problem and/or seems to the professor to capture the main strategy of the proof or problem. Even if you can't get very far at all, at least put down the few steps that you can do. You might earn 2 or 3 points for your efforts (the zero being reserved for students who haven't even started that problem). And hey, 2 or 3 points is 2 or 3 points.

★★★★★ 5-Star Tip

Think of your grader as a very rushed fellow human being. Make it as easy as possible for him or her to give you what you want—that shining A.

STRATEGY #15: Keep Up Your Stamina

Taking a long exam can sap your strength. It's easy to find yourself hitting the wall, like a marathoner toward the end of the run. But it's really important to avoid this. *Because more grades are won and lost in the last half hour of the final than at any other moment in the semester.* Many students blow their chances of getting top grades by letting their work fall apart in the very last few minutes of the test. On the other hand, many students break through the C/B barrier or the B/A divide by the work they do in those last minutes of the final.

Here's why. The last questions can loom particularly large because often the professor puts the hardest, longest, and highest-point-value questions at the end of the test (in part not to freak out students at the beginning of the final). Also your performance on the last questions is foremost in the professor's mind when he or she is making the grade decision. A bad finish is particularly unfortunate when an

exam ends up on the borderline and the professor has to decide which side of the fence it should come down on.

So be sure not to use up all your energy in the first part
of the test. Work in a measured fashion. Take brief mental
breaks from time to time, or even a minute to stretch if
you've got the space. And try to work up a second wind
when you come to the final stretch of the exam.

✓⁺ *EXTRA POINTER*

> If in spite of your best efforts you find yourself run-
> ning out of time, leave indications of how you planned
> to complete your answer. In a nondefensive way. Don't
> apologize for running out of time or for the brevity of
> your answer. Just say what your answer is, simply and
> directly. The CliffsNotes version. It's always better to
> get in even the briefest and most basic answer than to
> leave it out altogether because time is running out.
> Professors often grade exams looking for particular an-
> swers. As long as they find them in some form or
> other, even a stripped-down version, they credit you
> with the answer and give some (if not all) of the
> points. And, in the best case, the professor might not
> even notice that you were cut really short, and could
> have said much more if only you'd had the time. Espe-
> cially if you don't call attention to it yourself.

STRATEGY #16: Don't Panic Too Soon

The final is often the time when students are most prone to
panic. This can be disastrous because finals may have many
different parts and may often require a number of different
skills. One part might go better than others. And no one

thing will end up being fully determinative of the grade. So when you are in the process of taking a final, you often really have no idea how it is going to turn out. If you start monitoring things as you go along, it's quite likely that your sense is wrong. Don't let that process of evaluation send you into a panic. Because then it becomes a self-fulfilling prophecy.

Keep in mind that even if you encounter a question you can't answer, the answer could occur to you later in the test. On a math test, you might be unable to solve a problem or do a proof at one point in time, but then later in the test you might suddenly realize how to do it. When you still have plenty of time to go back. Maybe you have a short answer that requires you to know a certain date that you've forgotten, but later you remember it. Or maybe the first minutes after you read the essay topic you have no idea what an answer could be, but upon further thought the answer becomes obvious. Always remember that your test performance can turn on a dime.

Perhaps the best way to ward off panic is to use this mantra during your final: *It ain't over till it's over* (apologies again to Yogi Berra). This mantra works even better if you keep in mind that it ain't over even when you finish writing your exam. It's over only when the professor has to assign the final grades in the course. You never see or hear any of *that* part of the "over." But this is the time when a professor has a lot of discretion. To curve up the grades, even when the professor's policy is not to curve (a policy that goes by the wayside when the pattern of grades is lower than the professor can stomach). Or imagine this scenario: after reading all the exams, the professor realizes that he or she might not have done such a good job teaching some of the material that was tested. All of a sudden an essay that might have gotten a C+ magically turns into a B. And a B+ turns into an A.

This is not just something that could happen. It's something that *does* happen. All the time. Not everything that happens in a course is in your hands. So try to take it easy.

STRATEGY # 17: Don't Replay It Once It's Over

After you've finished one exam, move right on (in your mind, we mean) to the next one. Without looking back or second-guessing. Most often you don't know and can't really know how it turned out. And even if you do know—maybe you have the kind of professor who e-mails the grades the instant they're ready—it's all water under the bridge now. You still have three or four other tests to take.

Each one of these tests is independent of the others. Each affects your GPA pretty much the same amount. So there's no need to poison the well before the next test. Regroup and stay positive. If the first final didn't go all that well, that doesn't mean that the next one, and all the others, won't go better.

Review Session

What many students experience as their time of greatest fear, vulnerability, and in some cases desperation is in reality the opportunity for maximum empowerment. The chance to really take control, once and for all, of that mysterious process that is the college grading system. And to really take responsibility for that measure of how much you've learned—the grade. Here are 17 things you can do (neatly grouped into four categories) to make sure your semester ends with a bang:

1. **Improve your study techniques.** Extend exam week back to the last week of classes, and steer clear of "official" (but nonacademic) activities; go to review sessions or extra office hours (always with a question or two in hand); don't allot too much study time to courses you've got locked up; and be sure to make full use of all advance information available.

2. **Manage your *life* during exam period.** Use your own tried-and-true stress-management techniques; take it easy on substances and treat your body decently; and, when possible, shed commitments and enlist the help of well-wishers.

3. **Produce an excellent final exam.** View the exam as a final working session; make a plan and stick to it; construct your essay well, as if it were a paper; and make it easy for your grader to give you an A.

4. **Manage the *activity* that is the final.** Keep up your stamina, don't panic too soon, and once the exam's over, put it to bed.

Planning out a coherent studying and final-exam-taking strategy is the single best way to get an A in your college course. And to feel good—really good—about the course you've just nailed.

What Do Grades Mean in the End, Anyway?

If you've worked your way through all the chapters of this book—or even if you've just read one or two chapters on topics of interest to you—you're probably full of ideas about how the tips, techniques, strategies, and methods you've learned can be applied and customized to the courses *you're* taking. You know the five "grade-bearing moments" of the academic semester and, more important, you know exactly what to do to turn each moment into an opportunity for you to get that super-prized A. You're "charged up"—and well you should be, since you now understand getting good grades from the professors' perspective: the only ones who (besides you, now) really know.

And yet late at night you might find yourself wondering, "What do grades mean in the end, anyway? What's really the point of it all?"

For some, a transcript full of A's means a good shot at the law school, med school, or business school of their choice. And then a lucrative, fulfilling career to go with it. For others, it's a chance to get a good job after college—maybe in their home state or hometown (they'll ask about grades there, too, so all those A's will really pay off). For some students it's a chance to show their parents that the large investment they made in college was really worth it. And for others it's a chance to show their kids that Daddy or Mommy really *could* do it—that all the evenings at school really amounted to something, a set of A's.

But A's mean something more. They mean that when the professor picked your piece of work out of a stack of 70 others at 10 A.M. in his or her office—or at a Laundromat at Pico and 13th at 10:30 at night—it stood out as truly excellent. That for a brief 10 or 15 minutes there was a true meeting of minds between you and the teacher. Your professor or TA understood—really understood—what you were trying to say or do, and felt a sense of pride—that he or she would have been happy to have written the piece him or herself. A job well done.

If anything like this has ever happened to you—or if after reading this book you expect that it *will*—go into the next room and get yourself a purple or green felt-tip marker (if you're really stuck, a simple pen or pencil will do). Go ahead, we'll wait. Now fill in the box below and see what it really feels like.

Congratulations.

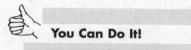

You Can Do It!

You now know how.

Join the Community

You're invited to join the *Professors' Guide* community. Maybe you have an idea for a tip, technique, strategy, or method that you've found helpful in getting A's. Maybe you have an interesting anecdote, experience, or story that you'd like to share with other college students. Maybe you'd like to spout off on some topic of interest about grading. Or maybe you want to disagree with something we suggested, and maybe you've got a better idea instead.

Join the community!

Go to our Web site: **www.professorsguide.com**. There you'll be able to "submit a tip," "submit an anecdote," and "submit an opinion." Send your thoughts to us!

The very best tips, stories, and opinions will be included in special *Student Input* boxes in the next edition of *Professors' Guide*. Headed by your name and school (if you give your permission).

We'd like to hear your input. *Really.*

Be a Pal

Maybe you know someone—or about 100 people—who could benefit from the ideas in the *Professors' Guide*. Hey, getting good grades isn't a zero-sum game.

So spread the word to your entire Facebook or MySpace list. Tell your friends that if they go to

www.professorsguide.com

they'll be able to read a sample chapter of our book, see our latest tips (free to you, too), get exciting offers, and join our interactive community. Then tell them that they owe you one.

ACKNOWLEDGMENTS

Many people have worked very hard to create the *Professors' Guide*. We'd like to thank especially:

★ Joseph Tessitore, president of Collins Publishing, and Richard Pine, (super-) agent at Inkwell Management. They understood our vision and knew how to make it happen.

★ Matthew Benjamin, our editor at Collins. He read the book countless times and made sure the tone never lagged and the tips were always authoritative, yet snarky.

★ David Christensen, Lynda Coon, Richard Lee, Ed McCann, and Ed Minar. Our best professor-friends, they spent endless hours talking with us about getting good grades, and offered many tips and anecdotes from their own experience.

★ Ana Maria Allessi, George Bick, Diane Burrowes, Brian Grogan, Phil Friedman, Libby Jordan, Jean Marie Kelly, Maggie Sivon, and Helen Song. Our partners at Collins, they worked tirelessly to get the *Professors' Guide* to you. And in good condition.

★ Michael Hyman, Chris Lederer, Sam Pinkus, and Celine Texier-Rose. They made dozens of business and strategic suggestions, all of which we did.

★ Haden Edwards, Andy Jett, and Al Reis. Masters of merchandising, they helped us refine our brand and discover what *Professors' Guide* was really about.

★ William Borchard, Michael Gross, and Richard Heller. They helped us with legal matters, both big and small, and taught us about management of content in the 21st century.

★ Sue Blanchard, Corrine O'Neill, and Ken Stout. Sue designed the male and female professor logos (they weren't intended to look like us), Corrine designed the boxes and icons and offered numerous page-layout suggestions, and Ken positioned the circled-A.

★ Jonah Hyman and Marianne Laouri. They made numerous proofreading suggestions that made the book better to read.

★ Martin Levin and Dan Weiss. They made many suggestions when the *Professors' Guide* was just a glint in our eyes; our voice grew stronger, as a result.

★ Yehuda Weg. He told us what to do when we didn't know.

And finally,

★ Our parents, Stanley and Marjorie Jacobs, and Arthur Hyman and Ruth Link-Salinger of blessed memory. Without them none of the other things would have happened—and you wouldn't be getting your A's just about now!

INDEX

recording references for, 237

scholarly sources for, 227–31

search engines for, 230

tips for doing, 224–38

vs. analytical paper, 205, 208–10

vs. report, 234–35, 265

See also paper(s); research

requirements, *See* course(s); language requirement

review session, 9, 75, 142–43, 315–16, 318

section leader, *See* teaching assistant (TA).

section meeting(s), 18, 21–22, 85, 307

activities in, 76, 117–18, 287, 294, 312–13

and exams, 117, 118, 130, 131, 136, 164, 176

and papers, 215, 226

note-taking in, 100, 133

value of, 73, 116–20

See also class participation

self-talk, positive, 291

skills, 41, 106, 130, 325

study, 56, 148, 172, 300, 314

writing, 50, 281, 325

standards, *See* grade(s).

stress, 65, 112, 120, 137, 146, 149–50, 177, 291, 293, 299, 320–21, 329–31

study environment, 292

study group, 108, 121, 149, 292, 307

study guide, 75, 132–33, 142, 143, 318

study questions, 110, 132–33, 142, 143–44, 145, 204, 316, 317, 325

syllabus, 16, 46, 68, 121, 301

and exams, 130–31, 138, 294

and first week of classes, 64, 80

and lecture notes, 90, 93

and office hours, 185

and reading, 110

and research papers, 225

decoding of, 67–77

taking notes about, 69, 76, 80

synthesis paper(s), 211

TA, *see* teaching assistant (TA),

taking stock, 121, 176, 276, 298

teaching assistant (TA), 22, 32, 116